Advanced Introduction to Cities

Elgar Advanced Introductions are stimulating and thoughtful introductions to major fields in the social sciences and law, expertly written by the world's leading scholars. Designed to be accessible yet rigorous, they offer concise and lucid surveys of the substantive and policy issues associated with discrete subject areas.

The aims of the series are two-fold: to pinpoint essential principles of a particular field, and to offer insights that stimulate critical thinking. By distilling the vast and often technical corpus of information on the subject into a concise and meaningful form, the books serve as accessible introductions for undergraduate and graduate students coming to the subject for the first time. Importantly, they also develop well-informed, nuanced critiques of the field that will challenge and extend the understanding of advanced students, scholars and policy-makers.

For a full list of titles in the series please see the back of the book. Recent titles in the series include:

Global Administration Law
Sabino Cassese

Housing Studies
William A.V. Clark

Global Sports Law
Stephen F. Ross

Public Policy
B. Guy Peters

Empirical Legal Research
Herbert M. Kritzer

Cities
Peter J. Taylor

Law and Entrepreneurship
Shubha Ghosh

Mobilities
Mimi Sheller

Technology Policy
Albert N. Link and James Cunningham

Urban Transport Planning
Kevin J. Krizek and David A. King

Advanced Introduction to
Cities

PETER J. TAYLOR
*Emeritus Professor of Geography,
Loughborough University and Northumbria University, UK*

Elgar Advanced Introductions

Edward Elgar
PUBLISHING

Cheltenham, UK • Northampton, MA, USA

© Peter J. Taylor 2021

All rights reserved. No part of this publication may be reproduced, stored in a retrieval system or transmitted in any form or by any means, electronic, mechanical or photocopying, recording, or otherwise without the prior permission of the publisher.

Published by
Edward Elgar Publishing Limited
The Lypiatts
15 Lansdown Road
Cheltenham
Glos GL50 2JA
UK

Edward Elgar Publishing, Inc.
William Pratt House
9 Dewey Court
Northampton
Massachusetts 01060
USA

A catalogue record for this book
is available from the British Library

Library of Congress Control Number: 2020950911

MIX
Paper from responsible sources
FSC® C013056

ISBN 978 1 83910 012 3 (cased)
ISBN 978 1 83910 013 0 (eBook)
ISBN 978 1 83910 014 7 (paperback)

Printed and bound in Great Britain by TJ Books Limited, Padstow, Cornwall

For Enid Taylor

What is the city, but the people?

William Shakespeare
(The Tragedy of Coriolanus, Act III, Scene 1)

Contents

Preface		ix
Preamble: academic literature on cities		xi
1	**City basics**	1
	City Insights A: Ramita Navai's Tehran	21
	City Insights B: Timothy Brook's Vermeer, Delft and a Global World	23
2	**Cities as the birth of civilizations**	26
	City Insights C: Brenna Hassett's Bioarchaeology	35
3	**Busy cities**	38
	City Insights D: Luiz Eduardo Soares' Rio de Janeiro	47
4	**Cities connected**	49
	City Insights E: T. H. Lloyd's German Hanse in England	58
5	**Demanding cities**	61
	City Insights F: William Cronon's Booming Chicago	69
6	**Divided cities**	72
	City Insights G: Hsiao-Hung Pai's Stories of China's Rural Migrants	80

7	**Cities in states**	**83**
	City Insights H: Anonymous' Berlin 1945	94
8	**Cities globalized**	**97**
	City Insights I: Ben Rawlence's Dadaab	111
9	**Cities in Nature**	**114**

Bibliographic notes and references 123
Index 129

Preface

Providing a short introduction to a subject as numerous and various, complex yet commonplace, and always contradictory as cities, is a tall order, or in more modern parlance, a challenge. From my perspective and given these high stakes, the only reason for agreeing to attempt this task is that it should be an enjoyable experience. And it has been. Putting together a wide range of descriptive arguments about cities turned out to be a fun task because it enabled me to scan a whole array of city characteristics, and pick and choose at my leisure. I have to admit that I've behaved like a greedy child in a candy shop! Cities are yummy as a subject to write about.

This is not to say that the selection of topics is random. There are many different ways of understanding cities and the path I have taken is a specific materialist one. By this I mean I start with the work done in cities, the raison d'être of their being. But these great concentrations of providing a living for residents should also enable these same people and their families to build meaningful lives. However, there is nothing certain about this. Thus, the cities I describe encompass the good, the bad and the wobbly. For many people cities are wonderful places to live, for others they are dangerous places to live, and there can be movements between these two states for many, many reasons. Hence wobbly is my way of describing this inherent uncertainty. Cities are elusive as a subject to write about.

My basic job has been to transfer my exuberance for cities to my readers. I hope I have done this in a coherent ordered manner that enables understanding to be gradually built up, whilst simultaneously providing interesting nuggets of information to reinforce the message. And the latter is simply that you cannot make sense of our social reality without regard to cities, something I believe to have been true for millennia. This is a hint that the range of cities I bring into my arguments is not limited to current urban worlds: I find ancient cities as fascinating as contemporary cities. But it is the latter that are to be indicted for travails facing us in the twenty-first century as we consume ourselves, potentially into oblivion, in, through and between our cities. Cities are vital as a subject to write about.

No writer is an island. After several decades of teaching, researching and writing about cities I have many debts to pay to others who have contributed to my understandings in numerous ways. My acknowledgements fall into three categories. First the basics: the wonderful writings of Fernand Braudel and Jane Jacobs. Never cross-referencing each other, much of my thinking has been about bringing together their insights about how cities work. Second there are a group of people with whom I have worked in projects and writing on cities who also double as friends: Mike Barke, Jon Beaverstock, Ben Derudder, James Faulconbridge, John Harrison, Michael Hoyler, Paul Knox, Rob Lang, Zachary Neil, Geoff O'Brien, Phil O'Keefe, Kathy Pain, Dennis Smith and Frank Witlox. Each will have influenced this book in many ways that they might not appreciate: they have certainly been supportive in getting things done but more importantly they each have brought different ideas and knowledge into my understanding of cities. Finally there are scholars who have written truly exceptional books about cities and inevitably their ideas are to be found in this book. Sometimes explicitly, more often implicitly, I have used their thoughts in developing my own text: Giovanni Arrighi, Neil Brenner, Manuel Castells, Edward Glaeser, Peter Hall, David Harvey, Doreen Massey, Jennifer Robinson, Saskia Sassen, Allen Scott, Ed Soja and Michael Storper. With minimal attributions in my text, I only hope I have done justice to their works.

And finally I do have to mention my wife Enid. The reader will find several interruptions in my text where I summarize books about cities from different positions, not scholarly but very insightful and always 'good reads'. Each of these précis has passed Enid's scrutiny to ensure my ruthless reductions maintain a clarity and coherence while still making sense. Hence this book's dedication.

Preamble: academic literature on cities

This book is an addition to a huge global industry: academic publishing's annual revenues are measured in many tens of billions of whichever major currency is applied. As support for learning, scholarship and research across the world this is a very positive thing. It is a response to massive increases in access to education providing opportunities for young people to enter professional job markets. But the sheer size of this supply and demand process has generated serious scale problems. The particular example of academic literature on cities is a clear case of insurmountable propagation of learning material.

Academic publishing deriving through the work of universities comes in two forms. First, there is the reporting of research findings and interpretations – new current knowledge in a field of study – that appears as articles in academic journals. Second, there are summaries and distillations of material from articles combined into academic books ranging from research monographs to student textbooks. This simple dual presentation of academic output worked well for about a century after the invention of the modern university in the decades spanning 1900. But no more: the academic publishing scale problem is manifest as more and more journals accompanied by larger and larger numbers of books in the twenty-first century. In this preamble I will illustrate this problem for the study of cities.

Academic journals were originally records of research by members of scholarly societies representing different disciplines with a country. These can still be found as very long rows of annual volumes on university library shelves across the world. However, in the second half of the twentieth century this simple production process changed as a whole new raft of journals was launched, some covering just parts of disciplines, others multi-disciplinary, and generally becoming more and more specialized. This is precisely what happened in what came to be known as 'urban studies'. Table 0.1 provides a list of journals with either 'city' or 'urban' in their titles ordered by the year of their first volume. It illustrates the result of this particular boom in English-language journals for my subject area. It appears that for those wanting to keep abreast of the latest research on cities in the anglophone world today there are 33 journals to be read.

The sequence of journals coming into being in Table 0.1 is typical of the expansion of this form of academic publishing. Beginning with general titles (e.g. *Urban Studies*) and sub-disciplinary offerings (e.g. *Journal of Urban Economics*), there is a general tendency to more and more specialized journals (e.g. *Sustainable Cities and Society*). Overall, the result is journals linking understanding of cities to a whole range of subjects – education, law, economics, history, architecture, geography, technology, health, environment, planning, design, agriculture – and in many cases there are more than one journal per subject. To provide an insight of where I fit into all this I can reveal I have published articles in just eight of these journals: *Urban Studies, International Journal of Urban and Regional Research, Journal of Urban Affairs, Urban Geography, Cities, Urban History, European Urban and Regional Studies* and *International Journal of Urban Sciences*.

However, many of my articles on cities have been published in other social science, geography and planning journals that do not have 'cities' or 'urban' in their titles. The point being made here is that the list in Table 0.1 does not constitute the whole universe of journals in which urban research is reported. But the bigger point is that even if the latter were the case, it would still be impossible to function as an active researcher and keep up with – read, digest and use – all the hundreds of articles published annually in these 33 journals and more. Houston, we have a problem.

Table 0.1 Academic journals on cities

Journal	Year of launch
Urban Studies	1964
Urban Affairs Review	1965
Urban Education	1965
The Urban Lawyer	1969
Regional Science and Urban Economics	1971
Landscape and Urban Planning	1973
Journal of Urban Economics	1974
Journal of Urban History	1974
International Journal of Urban and Regional Research	1976
Journal of Architecture and Urbanism	1976
Journal of Urban Affairs	1979
Urban Geography	1980
Urban Policy and Research	1982
Cities	1983
Environment and Urbanization	1989
Review of Urban and Regional Development Studies	1989
Urban Forum	1990
Urban History	1992
Journal of Urban Technology	1992
European Urban and Regional Studies	1994
City	1996
Journal of Urban Design	1996
International Journal of Urban Sciences	1997
Journal of Urban Health	1998
City and Community	2002
Urban Research and Practice	2008
City, Territory and Architecture	2009
Sustainable Cities and Society	2011

Table 0.1   Academic journals on cities (continued)

Journal	Year of launch
Urban, Planning and Transport Research	2013
Future Cities and Environment	2015
Urban Design International	2016
Urban Agriculture and Regional Food Systems	2016
Cities and Health	2017

Perhaps the second part of academic publishing, the books, can help here since an important purpose of these publications has been to build digestible-sized pieces of knowledge from researches reported in articles. Not so: recent proliferation in books is much greater than for journals. I have explored the current online catalogues of publishers whose books I regularly use and the results are reported in Table 0.2. Catalogues differ in structure, and therefore searching for numbers of books relevant to urban studies/city research does not necessarily produce consistent, comparable numbers, some being based on finding books using keywords, others counting books within specific urban sections. But the point being made here is simply the quantity of books available that perhaps should at least be perused if you aspire to keep up to date on what is being written about cities. With certainly more than 1,000 books waiting to be read from just these six anglophone publishers, we find ourselves again in an impossible situation even if we devoted all our waking hours to reading.

It was not always like this. In 1964 when *Urban Studies* first appeared it was feasible to keep up to date with academic publications featuring cities. This was important. Academics were expected to have a critical overview of their subject matter and this was the basis of their claim to be 'experts' in their field. It meant they could assure their academic peers, students, policymakers and publishers that they had a handle on all the key ideas, findings, concepts and theories in their knowledge area. I have a confession to make; in 2020, I do not meet such criteria to be an expert in urban studies/city research. I do not know, let alone read, many of the journals listed in Table 0.1, and although I have read numerous books on cities in the recent past I guess that these would make up far less than 1% of the catalogued books featured in Table 0.2. Houston, we still have a problem.

Table 0.2 Books in online catalogues from six major publishers, 2020

Publisher	Section/search	Number of books
Edward Elgar	Category: Urban and Regional Studies	522
Elsevier	Search: Urban	247
Elsevier	Search: City	282
Routledge	Textbooks: Urban Studies	199
Springer	Search: Urban	860
Springer	Search: City	558
Sage	Search Textbooks: Urban	95
Sage	Search Textbooks: City	88
Wiley	Search: Urban	425
Wiley	Search: City	356

There is a time-honoured way in which academics cope with an increasing volume of research knowledge: specialization. The sphere of knowledge for which expertise is claimed is reduced to manageable size. In the study of cities this is registered by adjectival descriptors in a diverse range of divisions of subject matter: geographically (e.g. focus on African cities), historically (e.g. medieval cities), functionally (e.g. global cities), methodologically (e.g. the science of cities), technologically (e.g. smart cities), environmentally (e.g. sustainable cities), and so on. There are bona fide experts in all these areas of urban studies. Thus we know far, far more about cities – empirical findings, conceptual insights, new ways of thinking – today than half a century ago. But, and it is a big but, in breaking up our subject matter into smaller pieces there is an ever-present danger of losing sight of the complex whole that is the essence of what makes a city. Specialization in any field is always susceptible to researchers being criticized for 'knowing more and more about less and less'. And for understanding cities this is an especially serious problem. Focusing on bits, however well conceived and well carried out, ultimately loses the human spark of what makes cities so different from all other settlements: cities are a clear case of 'the whole being more than the sum of its parts'. Thus it is necessary to periodically take a step back and think more holistically. This is how I interpret my job in writing this advanced introduction to cities.

1. City basics

Introduction: beyond demography

Irrespective of the number of journals and books devoted to them, cities will be very familiar to all readers of this book. In the twenty-first century most of us live in or near cities, and the rest rely largely on cities for many essential needs. As we go about our daily activities, cities are the backdrop that makes things happen. Loving them as fun and exciting, or loathing them as wicked and corrupt, cities are always there in our lives, in plain sight. Given such ubiquitous experience, what is there to introduce, advanced or otherwise? Just open a dictionary: in my *Oxford English Dictionary*, city is defined simply as 'a large town'. Job done?

No. We need something much more substantive than mere size to distinguish city from its urban twin, town, and other settlements. The basic point is that cities are dynamic; they operate in complex ways not replicated in other settlements. We can envisage a 'sleepy town' but a 'sleepy city' is simply no longer a city. Thus it is that cities not towns are seen as either fun and exciting or wicked and corrupt. It is true that city dynamism does owe something to population size – which is what the dictionary definition is perhaps implying – but being a city is so much more than just demography. The purpose of this chapter is to explore this special nature of cities and to set out the basics for use in all subsequent chapters.

Generic thinking on cities

I begin with what at face value is a curious observation: nobody disputes that Uruk in ancient Mesopotamia, Rome during its imperial height and Manchester in Britain's Industrial Revolution were each cities, indeed very important cities for their respective times and places. With historical intervals of millennia between them, they are immensely different in all manner of characteristics. But they must have something in common to be each identified indisputably as the same type of settlement, a city. In fact since the rise of Uruk 5,000 years ago this type of settlement has become increasingly commonplace, found across the world in myriad societies, presumably associated with similar arrays of economic, social, political and cultural functions. Similar enough, that is, to in some way 'qualify' to be designated cities in comparison to other contemporaneous 'lesser' settlements.

The ubiquitous nature of such identification is illustrated in Table 1.1. Simply using the Google translation service, I have been able to find renditions of 'city' in nearly 100 languages. Of course, these different languages, deriving from a wide variety of social contexts, will be signalling different settlement arrangements, but not different enough for standard translation as the English word city. The common denominator amongst the translations will be the recognition of settlements that are in some way special, superior to other settlement not thus designated. For instance, in medieval Europe settlements housing a cathedral were officially cities. In the USA's frontier phase many new settlements were designated 'city', as an aspiration that they would become special. Most failed to make it which left a difference in the meaning of city between 'English English' and 'American English'. However, today the meaning of city in the English language generally indicates special in terms of economic and cultural importance. (Small cathedral cities in Britain and 'failed cities' in the USA remain as linguistic anachronisms.) This is the 'modern' meaning; other civilizations will have alternative sets of criteria for identifying cities, key practices that make some of their major settlements special, aka cities.

Considering cities in this manner is a generic way of thinking. This approach can be applied to human institutions and practices that can be traced over long periods of time. Cities are just one of many examples: states, class conflict, religion, commerce, trading and core–periphery

Stad	المدينة	शहर	Grad	Град	城市	Ciutat	Město	Byen	ville	stadt
Πόλη	vil	עיר	शहर	lub zos	Város	Kota	città	都市	Mji	veng
Pilsētas	Miesto	tanàna	Bandar	Belt	شهر	Miasta	Cidade	Ar Dähnini
Города	aai	Mesto	Ciudad	Staden	oire	kolo	Şehir	Місто	شهر
Dinas	Noj city	alum	ከተማ	şəhər	горад	শহর	syudad	mzinda
cità	urbo	lungsod	stêd	cidade	ქალაქი	શહેર	gari	kūlanakauhale
obodo	cathrach	kutha	ନଗର	кала	bajar	шаары	ധഠയ	urbs	pilsēta
miestas	ಊರು	taone	хот	by	ਸ਼ਹਿਰ	oraş	bhaile-mòr	toropo	guta
magaalada	kota	шахр	Kent	shahar	thành phố	umzi	ilu	idolobha

Table 1.1 The word 'city' in multiple languages

dependency have all been treated in this manner. Thus none of the above was invented in our modern civilization but all are to be found in a modern form. The basic thesis is that each institution is a generic formulation available for moulding into different realities in different social contexts. The exception is the first creation of an institution; such origin processes are fiercely debated in all cases. For cities, creation has been traditionally linked to the birth of prime civilizations so that they have several origin places – Mesopotamia (Iraq), Egypt, China, India, Mexico and Peru constitute the customary list. It is from these and other multiple beginnings that cities have come to be found across the world as indicated in Table 1.1. The key question arising from this generic approach is what is it that all cities have in common – what in general makes them cities?

Generic understanding of cities is hugely dependent on acute abstraction, a peeling back of multiple characteristics of known cities to reveal a basic process that defines a nature of cities. This can be done in different ways depending on the purpose of the study. It is not necessary to provide a comprehensive review here but brief mentions of some influential examples will help in placing my preferred approach in context.

Two mid-twentieth-century examples stand out. In the first half of the twentieth century the Department of Sociology at the University of Chicago pioneered what is commonly viewed as the first intensive attempt to understand the complexity of cities. This 'urban ecology' culminated in Louis Wirth (1938) defining 'urbanism as a way of life' consisting of three key characteristics that, in combination, other settlements could not provide: population size (providing anonymity), density (enabling specialization) and heterogeneity (stimulating experience). Collectively these provided an ecology that generated both the benefits and ills of urban life. Scott and Storper (2014) continue this American take on cities, updating terminology while setting it within broader debates, by focusing on a combination of agglomeration and associated land use processes as distinguishing the 'urban' from wider society. Meanwhile in archaeology, identifying – and therefore defining – cities had been a central activity in excavations culminating in Gordon Childe's (1950) identification of the 'urban revolution'. He provided ten criteria for distinguishing cities from other settlements, centred on the creation of a social surplus. This derived from population size and density with a division of labour enabling monumental public buildings housing a bureaucracy operating accountancy and writing. Smith (2016) has recently increased this to 21 archaeological urban attributes in a sophisticated comparative study of ancient cities. Generic approaches to cities continue to be developed in very contrasting ways ranging from Amin and Thrift's (2017) tangle of networks in a 'seeing' and 'thinking' city to Batty's (2013) 'new science of cities' using 'power laws' of city distributions.

Thus, I am not short of choice in finding a suitable generic approach: all of the above have their merits based upon sound understandings of cities. Elements of each are infused in my approach. However, I go back to basics. I start with a very obvious observation: in order to become different from other settlements – more than simply quantitatively bigger but in some sense higher in quality – cities have to develop a recognizable pre-eminence. Thus I argue that what makes cities is the nature of their growth. Understanding cities cannot be separated from explanation of city growth beyond population enlargement.

This thinking leads towards a different academic genealogy, a more explicit economic take on urban growth. Alfred Marshall (1890), one of the founders of modern economics, coined the term 'industrial district' to indicate places where ways and means of production were informally shared as if existing 'in the air', as he famously termed it. His pioneering work provided an alternative economic focus to mainstream modern modelling of national economies and has produced various strands of regional economics and economic geography. However, in fact he was initially referring to the nineteenth-century concentrated growth of cotton and steel manufacture around Manchester and Sheffield respectively. But these were plainly city processes, with city being geographically viewed as a broader integrated economic city-region rather than just a bounded local administrative area. This notion of cities as economies has been developed in the second half of the twentieth century by Jane Jacobs (1970, 1984). It brings together much previous generic thinking, while providing a more unified materialist grounding, and culminates in viewing cities in ecological terms (Jacobs 2000) harking back to the Chicago school. Its key advantage is the ease with which it can be deployed across the whole historical and geographical range of cities. My generic approach builds upon Jacobs' legacy.

Cities as generic can be encapsulated in the following five defining features.

1.   Cities as process

This is our starting point revealed above. At first it may appear a curious idea – we experience cities as places; all cities have their iconic images as skylines of buildings. But these are just episodes in an ongoing story, transient representations of a city continually unfolding; every city is always a work in progress. And this work is immensely complex: the sum of myriad individual movements of people making a living to make a life that creates and recreates a city alongside other cities. Cities are different from all other settlements through their inherent economic and social dynamism.

This defining process of cities is not a random mass of interactions; its complexity is systemic. Every city is a complex whole that generates huge social changes, notably economic development. Cities are always growing in myriad little ways so that change is endemic. Each adopted change throws up new possibilities; inevitably there will also be problems, for which solutions have to be found. This is not a smooth evolutionary process but rather one of trial and error. And with solutions, cities are where new and old ways of doing things coexist in uneasy tensions of simultaneous advance and decline. This is called organized complexity, city as a perennial development.

EXAMPLE: Birth of the Metro
The systemic nature of cities is most obviously expressed in the infrastructure enabling movement of people and commodities. Here is the problem: in periods of rapid city growth old transport systems become increasingly impracticable and have to be replaced. Such was the situation during the nineteenth century in Western Europe and the USA where numerous cities were growing to a million population and more. The primary means of transport involved horses, many millions of them. The largest city was London where tens of thousands of horses pulled carriages, carts, cabs, buses and trams creating an inefficient (chaotic congested streets) and very unhealthy (many tons of dung on streets) transport system. And here is the solution: London responded with the first underground railway in 1863, but the steam engines made for a very uncomfortable journey. New York started an overhead railway in 1868 based upon stationary steam engines and cables, but to limited success. It was the use of electricity to power the trains that finally heralded a change in the transport system. In 1890 London's electric underground system began that continues to this day. In the next two decades electric trains, both underground and overhead, were built in cities across Europe (Paris, Berlin, Budapest, Glasgow, Athens) and the USA (New York, Chicago, Boston, Philadelphia), and the numbers of horses in city streets began their rapid decline to all but disappear from the streets of cities in the twentieth century. Today there are nearly 200 metro systems in cities across all continents, and growing.

City as process – organized complexity – is a meta-construct encompassing a cacophony of urban processes. The other four city defining features are such key processes, each contributing to what makes a city a city: how city development is created, shared, maintained and threatened.

2. Agglomeration of activities

This is the process out of which a city's unique capacity for innovation results. It is a product of both the size and diversity of cities that distinguishes them from all other settlements. Glaeser (2011) refers to urban dwellers being 'smarter' people, not innately so as individuals, but collectively because they live in an enabling social context alongside multiple other 'smart' people. This advantage is specified in economic terms as an externality, meaning a benefit outside ordinary market returns. Thus projects by an entrepreneur working in a city are more likely to succeed than those of a rival working in a non-city environment. This effect is termed an agglomeration externality.

Here are some of the things that provide an advantage to city businesses:

- People working in the same economic sector from whom you can learn as well as compete
- A wider range of people from other related sectors from whom you can learn
- A wide range-skilled and unskilled labour readily available
- A range of multiple sources for procuring capital
- Materials/parts available from multiple other producers as and when needed
- Local demand for a product and the possibility of developing new demand
- Specific trade organizations and similar institutions for support and advice
- Anonymity of activities providing opportunities for illicit activities, including criminal activities.

Taken together these operate as a sort of support system for economic development. But we don't have to stop here: agglomeration processes can be found in other urban activities. Innovations in political, social and cultural practices create city developments that are equally as important as economic change in making cities. And they are truly generic: writing this book, I am heir to the agglomeration of scribes in Uruk, and other Sumerian cities, who invented writing some 5,000 years ago.

EXAMPLE: Hollywood

This is undoubtedly the most well-known industrial cluster in the world. Beginning in the 1910s, by the 1930s Hollywood was the world capital of the motion picture industry. As Allen Scott (2008) tells it the district has since developed into a broader entertainment industry cluster with television production and music recording adding extra celebrity kudos to the name. It operates as a mix of competition and cooperation through a combination of large and small companies. There is a perennial kaleidoscope of fluid networks of deals and projects in multiple joint ventures. Larger firms take charge of finance, production and distribution while bringing in smaller specialist firms to input work on things like artwork, editing, programming, videos and content research. In addition there are key organizations such as the Academy of Motion Picture Arts and Sciences and key publications such as *Billboard* music magazine. These all add up to agglomeration externalities unrivalled in the sector worldwide.

3.   Connectivity of activities

The main focus of urban studies of the external relations of cities has traditionally been on their hinterlands. But no city ever developed by relying on just its immediate surroundings. Cities come in groups, never alone. And it is the connections between cities that are crucial to a city's development. The power of their agglomerations is nourished, supported and replenished from outside: new people, new ideas and new demands, operating through larger network mixes of competition and cooperation. For instance, London and New York have immensely important financial clusters but they would be nothing without the myriad financial links between each of them and with multiple other cities across the world. London and New York also have the most external financial connections: this city advantage is called a network externality.

The most productive way of understanding network externalities is through Manuel Castells (1996) ideas of social space, the way in which humans make and use space. He argues that social space is developed in two distinctive forms: spaces of places and spaces of flows. The former is typically represented on maps as a patchwork world, for instance showing urban and rural places or, that most well known of cartographic images, the world map of countries. But each space of place is dependent on spaces of flows – of people, of materials, of ideas – in its creation and subsequent changes. Castells uses this distinction to claim that contemporary society is particularly marked by its intensification of spaces of flows and thereby can be labelled 'network society'. This may be, but his two concepts of social space are generic and spaces of flows are fundamental to understanding all cities, they are the stuff of network externalities, past and present.

Spaces of flows are constituted by three layers of interactions.

1. Enabling infrastructures. Historically these have been physical connections between cities: earliest civilizations along rivers, later cities through seaports, canals, roads, railways and airways. In the modern world these connections have been joined by virtual infrastructures: telegraph in the nineteenth century, telephone from the early twenti-

eth century culminating in today's global internet. This layer of flows provides the operational hubs of the network.
2. Social interactions. Traditionally these connections were made largely by people in the business of trade and finance. Today this has grown massively with transnational corporations operating through multiple cities throughout the world, the most explicit expression of Castells' network society. These business links are augmented with a vast array of other activities: political (UN agencies in New York, Geneva, Rome, Vienna, etc.), social (humanitarian organizations in London, Nairobi, Brussels, Bangkok, etc.), professional and creative organizations (conferences, meetings, festivals and symposiums in New York, Los Angeles, Paris, etc.). All result in continuous physical and virtual circulations of people and ideas, the nodes of the network as increasingly cosmopolitan milieus.
3. Elite separation. There are the spaces of elites, very wealthy cosmopolitans living in distinctive enclaves interconnected by micro-networks through which practical and cultural activities are conducted. The rest of society is priced out of these rich oases – residential and leisure – while the elites' micro-networks maintain strategic links to the other places and flows, to the rest of society. This layer of social flows will be specifically investigated in Chapter 8 where we consider globalized cities.

Castells' global network society operates through a mixture of hubs, nodes and enclaves in multiple networks of cities.

Network externality is relatively simple to envisage but harder to pin down compared to agglomeration externality. We can understand that a firm operating in London with its massive connectivity – both physical infrastructure through its airports and economically through its financial services – has a benefit over a similar firm operating in a smaller city, but these benefits are perforce intimately linked to agglomeration. The face-to-face interactions resulting from external links are aligned with the agglomeration advantages listed previously, except operating on a wider scale. They are treated here, and nearly always in the literature, as separate processes but the agglomeration externalities and network externalities are two sides, literally – internal and external - of the same process, what we might refer to as 'whole city externality'. Keeping them separate is a pragmatic decision for exposition of a multifaceted arrangement, a necessary choice typical when describing complex holistic systems.

EXAMPLE: American Civil War

From the 1840s onward railways heralded a revolution in connections between cities. In the USA this coincided with increasing political tensions between North and South culminating in the civil war from 1861–65. The pattern of railways in 1860 demonstrates this to be a very unequal military contest logistically: connectivity between cities in the North was totally different to that in the South. In the former there were densely interconnected northeast seaboard cities (Boston, New York, Philadelphia, Baltimore), linked to dense northwest central interconnections (Detroit, Pittsburgh, Cincinnati, Chicago) creating an increasingly integrated North of linked city agglomerations. In contrast railways in the South were few and formed simple 'tree' patterns, single routes inward from the coast such as the Savannah-Atlanta-Chattanooga line. There are many reasons why the North prevailed in the war but differences between cities and their connectivity are immensely significant. And it is this network of the North's cities that develops into the American Manufacturing Belt, powerhouse of the US economy for the next century including being the victorious industrial war machine in two world wars.

4. Projection of power

No social change comes about without an application of power in some form or pattern. Words like evolution and progress tend to imply that change, indeed transformations, happen simply as an ongoing feature of society. But power relations are ubiquitous within society. Power is also one of those concepts that appear to be both simple and complex at the same time. So what does it mean to say that a city is a powerful social force, or to say Paris is more powerful than Lyons? Paris is the bigger of the two cities, Paris is the capital city of France, both are relevant, but Paris' perceived forceful prowess is much more than these two facts. If we change focus to Italy and ask which is more powerful Milan or Rome, we can begin to see a way forward: Milan is the leading economic centre and Rome as the capital is the political centre of the country. It depends what you mean by power? The question is: which power? With respect to cities there are four basic answers.

Command power is the usual way we think about power. It is having direct power over social change, an instrument power – the power to do things. Thus states have power to change matters in their territory and this is formally conducted in their capital city. More generally, mayors or civic leaders have similar but lesser powers over the area of their city. But command power need not be boundary limited. In the headquarters of large corporations decisions are made that affect all their work activities wherever they take place. Thus in cities housing many corporate HQs, like New York, Chicago or Beijing, there is the power to close a factory in city X, open a new factory in city Y, while simultaneously expanding work in city Z. Today, cities where such global strategies are devised and acted upon are called Control and Command Centres.

Networked power is a more subtle form of power that derives from the everyday work in cities. It is the power that holds networks together, the work that facilitates and develops links within and between cities. This capacity to shape a situation relates to the agglomeration and clusters of cities where a wide range of associations are negotiated and mediated in a collaborative process of leverages that build operational bridges both within cities and across cities. This is how financial centre clusters work through a range of professional inputs: financial analysts, fund managers, wealth advisers, lawyers, accountants, etc. Thus do cities like New York, London, Frankfurt and Tokyo have power within global financial markets.

Structural power is a more diffuse and embedded form of power. It operates in the everyday workings of unequal social conditions where dependency is the crucial relation. This is the situation between a city and its hinterland, the latter being dependent. What happens to the city – growth, decline – will automatically affect the hinterland in a situation outside its control. This type of relation plays itself out at much larger scales where regions specializing in specific food products or raw materials are dependent on faraway markets. For instance, futures markets in Chicago – betting on future prices of commodities – affect the livelihoods of farmers in South America but they are powerless in this price-making exercise.

Ecological power is the inner strength of a city's organized complexity; it is a sort of combination of the other three powers. As a social ecology there are innate processes that make the city, and these can vary over time as the city waxes and wanes. But collectively, cities have proven to be immensely powerful ecologies: they have grown and grown over several millennia to such an extent that today they both dominate our species and at the same time endanger the global ecology that supports all species. This capacity to generate such awesome power for social change is proving very difficult to deal with in the twenty-first century. This problem is engaged with directly in the final chapter.

EXAMPLE: Hong Kong

In terms of simple command power Hong Kong is conspicuous by its dearth. It became a colony of the UK in 1842, Britain's trading foothold in southern China, and kept this political status until 1997 when it returned to China with ultimate political control transferring from London to Beijing, aka from imperial command to communist command. Despite this political pedigree Hong Kong has been and continues to be a hugely successful city as other forms of power have played out. It is an exemplary example of networked power. Hong Kong's prime role has developed as commercial gateway between China and the West. This involves agglomeration of both practical knowledge and tacit knowledge on how to carry out business between these two different political economies. Thus financial advisers will know how to navigate Beijing's rules and Shanghai's commerce; lawyers will know how to accommodate Chinese contract law with the laws of other countries in drawing up international deals. It is one of the most important operational bridges in the global economy today. Without command power, nevertheless it is one of the most powerful cities in the world.

5. Relations with states

As noted above, the power of cities exists alongside political power, deployed through the might of warlords, states and empires. Thus through most of their history cities have operated in a world of contrasting power logics: alongside their networked power premised on mutual interests between actors in spaces of flows, there is the state's command power leading to just winners and losers in spaces of places.

Capital cities are the most obvious example of city/state relations; as states take on more functions they require more complex organization. State bureaucracies grow within capital cities and can benefit from an agglomeration effect. As well as diplomatic activities and embassies, these capital political agglomerations include specialist think tanks and multiple lobbyists, plus specialist services such as language translation and security/surveillance, all added to government representative and administrative functions. But the scope of this work remains territorial, matters of state, and is only a minor input into overall inter-city relations. Thus, where a state creates a city to become its capital – Washington DC, Canberra, Brasilia and Abuja are important examples – the 'whole city externality' is compromised. City growth depends on the state, aka command power. For instance, Washington DC was a minor US city for most of its history and only became a major city with the rise of the USA as a leading world state in the second half of the twentieth century.

More generally, city/state relations play out in three main ways.

(1) State domination

  (a) States have formal power over cities, they are incorporated into territorial administration structures where the spatial relations are hierarchical rather than networked.
  (b) This dominance is most overtly shown in language: it is the state and not the city that has the naming rights of the city. This simple but highly emblematic power typically follows administrative reorganizations, wars and revolutions.
  (c) When capturing another state's city or putting down a rebellious home city, a state lays sieges, which is to say they prevent network externalities operating and thereby also close down the agglomeration externality.

(2) City/state accommodation

  (a) But the state's command power is a relatively crude instrument in comparison to the complexity of cities. Thus cities carry on through the rises and falls of states: today the vast majority of cities are much older than the states that currently house them.
  (b) Cities gain from the social order that states impose such as rule of law for transactions and safety in transit between cities.
  (c) There is a wide range of other positive city/state relations, different degrees of city fiscal dependence on state largesse.

(3) City domination

  (a) Since states and their bureaucracies are not able to accommodate the complexities of cities, their attempts to stamp their order on the city through twentieth-century spatial planning

emphasizing boundaries (zoning) and land control (urban limits) has largely failed.
(b) In the command economies of the twentieth century, communist states imposed severe state controls on city growth thereby undermining their economic growth. This created a political crisis in both the Soviet Union and communist China that came to a head in the 1980s: the Soviet Union disintegrated, China transformed.
(c) Illicit city operations have grown with the modern city; in the USA the so-called 'war on drugs' has been totally undermined by the demand for drugs in US cities with the external effect of subverting supply states, even one as large as Mexico.

Today the issue of city/state relations is at the heart of debates about economic globalization as large corporations use their own combinations of city locations to structure their transactions for minimizing their tax bills. This strategic combining of agglomeration and network externalities poses a huge threat to the territorial power of states. Strangely, city/state relations are a relatively under-researched topic in urban studies.

EXAMPLE: Wroclaw, etc.

Naming rights to cities have been mentioned above and here I provide what is perhaps the most extreme example. According to Davies and Moorhouse (2003), working backwards the capital city of the Silesia region has only been called Wroclaw since 1945 after its province was incorporated into modern Poland. Before then back to 1871 it was German Breslau, before then back to 1741 it was Prussian Bresslau, before then back to 1526 it was Habsburg Presslaw, before then back to 1335 it was Bohemian Vretslav, and before then back to 1000 it appears in a chronicle as Wrotizla. Of course, these name changes reflect the turbulent political history of Central Europe where resource-rich Silesia lay between various Polish, German and Czech political entities. But the key point is that this important city on the River Oder survived, building agglomerations and using connectivity advantages through changing populations.

These five defining features of cities are described in more detail, including their crucial interweaving, in eight substantive chapters that follow: in civilizations, busy-ness, connections, demanding, divided, in states, globalized and in nature. Each chapter ends with a 'concluding supplement' rather than a 'conclusion'. Typically the latter operate as summaries but these tend to preclude forward discussion. Instead my supplementary offerings aim to open up chapter subjects to new horizons, to stimulate thinking outside the box, aka the chapter.

Specifics of individual cities

Generic thinking is not enough. I consider it a necessary step for an advanced introduction to cities but it encompasses an inherent limitation. Accompanying the common reality of city complexity there is an equally resonant reality that all cities are unique. Not to recognize the latter would be a serious drawback for any introduction to cities. I began the discussion of generic thinking with the example of industrial Manchester, imperial Rome and ancient Uruk, each recognized as cities despite being so very different. This statement makes its point for the generics argument very strongly but actually differences are true, albeit to lesser degrees, for any other trio of cities, say Dallas, Houston and San Antonio in Texas.

City uniqueness is not trivial, even from a generic perspective. Quite the opposite: it is essential for the second and third defining feature of cities as enumerated above. Quite simply, a hypothetical grouping of multiple identical cities generates no interconnections. In such a situation there would be no need for trade between cities, no migration given that there would be no reason to move, and diffusion of ideas makes no sense. Without the stimulus of network externalities the existing agglomerations would lose their vitality in an absence of cosmopolitanism. Thus differences between cities are a necessary condition of the inherent urban dynamism that the generic approach is designed to explicate.

There is, however, another equally important reason for considering the specifics of cities alongside their generic properties. Multiplicity of complexes equals immense variety. It is within the idiosyncratic natures of particular cities that we find the true excitement and attraction that marks out cities – traditionally 'streets paved with gold', later their 'bright lights'. In generics these equate to 'opportunities' but the appeal of cities is so much more than conveyed by such a staid concept. This vital attraction is an additional coating to the dynamism of cities that generic thinking has defined. The key characteristic is unpredictability, every experience is different: cities are infinitely diverse environments, simultaneously exhilarating and dangerous. As such cities have played important roles in many novels; more than a backcloth to the action, they often appear almost as characters in their own right. Analysis of such nuanced fiction is a reach too far for this author – critical literature review is way beyond my skill set. This is an invitation to go where I have feared to tread. (However, I do have a weakly informed impression that science fiction in particular gets cities wrong.) But these reflections do suggest that for understanding cities we might look beyond the usual suspects – social scientists carefully doing their urban studies – to explore the specifics of individual cities.

Back to non-fiction texts, I have long admired the research of investigative journalists whose work often strongly features cities. Their best reporting occurs where they unravel the intricacies of cities with which they have a close personal relation. Sometimes part-memoir or returning to cultural family roots, they expose the complexities of 'their city', its idiosyncrasies laid bare for better and for worse. And from a different angle there are historians and archaeologists with a specific urban expertise who have devoted detailed research effort to understanding particular past cities. The kinds of evidence they deal with to reconstruct cities – city archives, city excavation sites – provide distinctive takes on cities, and often, crucially, as unique holistic developments through linking economic, cultural and political texts and artefacts. All this adds up to a variegated rich seam of scholarship that can be harnessed to my objective of bringing in city specifics to further generic understanding.

The basic position I take is that a city's uniqueness is the distinctive way in which the five generic features and their interactions have unfolded; the result is the creation of a wonderful array of urban differences. However, adding concern for the uniqueness of cities to a generic understanding of cities is not straightforward because we are dealing with different types of knowledge. This does not mean that each cannot be used to enhance what the other form of comprehension provides. But the result is not an integration of the generic and specific; we have to make do with intermingling them. The generic knowledge provides an underlying structure to the text, a sequence of chapters with themes that elaborate on the five defining characteristics of cities and their interrelations. The specific knowledge is presented as a series of additional short vignettes called 'City Insights', each drawing on a selected source that provides exceptional intuitions from and about their city or cities. These are scattered throughout the text to be pondered between reading the chapters. In effect, through this juxtapositioning I am providing two 'reality checks', with generic ideas facing unique happenings, and specific musings alongside broader understanding. The first two City Insights appear at the end of this chapter, one by a journalist and one by an historian, thus providing a first glimpse of these alternative knowledges. But in addition, every reader should begin making connections between the different knowledges by reassessing their own experiences of cities, especially 'their city'.

Preparatory provisos

As an advanced introduction this has to be an interactive text steering readers' existing knowledges of cities. Steering where? Hopefully towards a deeper understanding, sometimes unthinking of received wisdoms, always critically assessing what and how we know. This means that provisos about the content of subsequent substantive chapters have to be part of this chapter's introduction to city basics. Often in the argument developed in a textbook the crucial riders come at the end, caveats to prepare the reader for the fact that the world is always much messier than any systematic knowledge of it. Thus all my arguments about cities that follow are far too neat, they wrap complicated realities into meaningful but limited chunks. The City Insights are intended to lessen this intrinsic problem, but by their very nature they themselves have limited reaches. I have identified four key provisos that readers should be critically aware of whilst reading this text.

The first thing to appreciate about this text is that it is not strictly about 'urban studies'. The searching for journals and books to gauge the size of the literature in the Preamble used two terms (and their derivatives): urban and city. Often used interchangeably, they actually imply alternative ways of thinking. The concept of urban is essentially place based; it contrasts with a different type of place, rural; and, in the twentieth century, suburban. Defined as place, urban can be measured as urbanization, which equals the proportion of a country's population that live in urban areas. As a variable it can then be used in analyses with other variables to describe differences between countries. The concept of city is much more nuanced. It is understood as a place but in a functional way that means it cannot be simply bounded. It is located within spaces of flows, where boundaries are found and treated as obstacles and opportunities, but they do not define cities. Cities are active, multiple vital nodes. The four provisos stem from my choice of focusing on city as process over urban and urbanization.

This choice is not a mere play on words, it identifies a very different approach to the subject with important consequences. Studying the 'urban' implies that cities are reflections of the societies that contain them. This is often clearly marked by the adjective used to describe particular types. For instance, the concepts 'capitalist city' or 'Indian city' both, in their very different ways, assume the said cities are the result of a set of wider social processes. In contrast, viewing the city in a space of flows implies it is cities that make their societies; it is the latter that is dependent. This opens up study of cities to all aspects of society since it instigates this wider social whole. Urban studies covers only things urban by definition, thereby typically missing out other topics about things or processes in non-urban places or at non-urban scales. This is by no means a hard and fast distinction but by being on the city side of the divide, this book can take advantage of the freedom to go beyond standard urban studies, as indicated by including the power of cities and city/state relations as basic defining features of cities above. The proviso here is that the reader should be aware that this book is premised on social theory that is overtly out of the ordinary, back to front as many would see it.

The second proviso follows from choice of city over urban from another angle: the latter concept plays down the dynamism of cities, their vitality. And yet their very real success can be plausibly interpreted as phenomenal, culminating in today's global urban world. But in such a context, a focus on city can become a naïve celebration of the city. This is implicit in the definition of the city process as urban growth. Cities are distinguished from other settlements by their ability to grow faster and larger. Some cities stop growing, some disappear, but the focus tends to be on the good times, their growth episodes. In fact focus on contemporary cities in decline is typically about restarting the growth to become a 'proper city' again. This is, of course, all about how success is defined. So beware of ignoring the victims of every city success story: cities have never been idylls of prosperity for all, they each have their dark sides, partly hidden but with poverty always in plain sight within the city and beyond.

Focusing on city creates a difficult linguistic proviso. This is about something that occurs frequently in writings on cities, is widely understood as misleading, but equally is hard to avoid. I am referring to reification, turning a process into a thing. I have argued that cities are always a work in progress but this is not always easy to present, or at least maintain, when cities are mentioned multiple times in a text. It is a common difficulty with our language where sentences pivot around verbs using subjects and objects. It portrays a simple cause-and-effect world of insulated functions. Thus we are prone to say a city does this and that as if it were a singular actor rather than the complex whole I am introducing in this book. 'Manchester led the Industrial Revolution' or 'Detroit made cars like no other city' are both reifications. Of course, neither city did anything of the sort; cities don't 'do' such things. In both cases it was a process that unfolded in part of the city, which came to dominate, enabling reification to appear credible. Reference to cities 'doing' something may be reasonable in policy circles with reference to what a city government is carrying out – but best use 'city hall' or 'local government' – but is wholly inappropriate with reference to the complex unfolding that is city process. But our language is what it is and reification makes for easier reading, which is itself a good thing. We can interpret city reification as convenient shorthand for complex processes that, hopefully, are made clear by the context in which a sentence is embedded. But whatever, the key point, for both writer and reader, is to never fall into the trap of thinking cities are simple.

The final proviso concerns the difficult issue of a scholar's positionality: who we are and where we come from affects what and how we study. This has been a severe problem with respect to understanding cities because particular cities have commonly been promoted as especially important, a sort of 'all roads lead to Rome' or Paris, or Vienna, or New York, or Chicago, or Los Angeles, or wherever the promoter comes from, usually a particularly vibrant 'modern' city of the 'West'. So where does this leave my claim to have outlined a generic approach to cities above? And it is not enough my adding specifics, they are chosen by me and interpreted by me! There are three critical predispositions in my thinking: an emphasis on process – to cope with complexity; an explicit start from a materialist position – we all have to eat; and a strong historical orientation – a distain for 'now-ists' who can only see their own experienced present as meaningful. Unfortunately for the last there are always fresh interpretations arising as each new generation makes its mark. This has made me wary of declarations of 'post-something' scholarship because it begs the question of what comes next. Post-something implies an end of history unless you are willing to countenance a series of 'post-anythings', heaven forbid! Positionality can best be handled through honesty and modesty, but should never be ignored.

These four provisos should be kept at the back of the reader's mind as each of the following chapters spill their beans. They will overtly come back into focus in the final chapter.

City Insights A: Ramita Navai's Tehran

'In order to live in Tehran you have to lie' is a quotation taken from Ramita Navai's (2014) *City of Lies* (p. xiii) in her search for truth in Tehran. She was born in the city into a family of the Iranian political class who fled the country after the Iranian Revolution in 1979. Growing up in Britain, she became an international journalist and returned to Tehran in that capacity in the 2000s. Using multiple interviews with a wide range of citizens, her book is a telling of eight contrasting stories of people's 'love, sex and death' in navigating life under a repressive religious state. She describes herself as 'an outsider yet still a Tehrani' (p. 280).

Navai is not saying her subjects are congenital liars, they lie to survive: 'morals don't come into it' (p. xiii). This is a pervasive feature of life in the city, encompassing all classes and all levels of religious adherence. The Islamic Republic of Iran has created a state apparatus that combines with a powerful religious identity to generate a perpetual atmosphere of threat. Therefore, risk analysis is an integral part of navigating a life for everybody living in this great city; corruption is endemic as the prime mitigation tool. This leads to perhaps an unexpected outcome: politics is pervasive, invading 'conversations in every corner of the city' allowing people to believe that 'they are not powerless spectators' (p. 46). But the devil is in the detail, and there is a lot of detail in a city of over 8 million people, far too much for any oppressive state to master.

Navai begins by presenting a simple geography of the city centred on the Val Asr, a long road running from the station in the south towards the mountain foothills in the north. Created as part of an exercise in modernizing the city in the 1920s, the road connects what Navai calls two different countries (p. 281): poor working-class districts in the south and 'the bubble of north Tehran' that includes 'the Chelseas, Knightbridges and Mayfairs of Tehran'. The latter is 'the new Tehran, where tradition and class are blended together and trumped by money' (p. 257) but still not immune from the state – misdirected bribes lead to a rich family's wedding party raided for its decadent opulence. But it is in the south where 'the dark corners of Tehran' (p. 281) are to be found. These come in two forms: devout working families who self-police for morals in their neighbourhoods, and criminal zones jointly run by gangs and local police.

There are gradations along Val Asr, material success is announced by moving further north, but aspirations are particularly dangerous in the middle. Here the morality police ply their trade of lashings or bribes targeting improperly dressed young women in particular (pp. 83–4). It is where the northern bubble is frequently burst, especially by the younger generation of the rich going 'downtown' like their ilk worldwide: it is 'where Tehran still has heart and vigour' (p. 262). Here there is a large shopping centre full of Western stores – conspicuous consumption is rife (p. 154) – plus a vibrant entertainment industry, part illicit, part very careful.

But Tehran also has its own distinctive agglomeration processes. There is an industry of Islamic divination providing expert advice on important matters of life ranging from property and inheritance to adultery and divorce. These are services provided for a fee, sometimes sizeable, by mullahs, often with assistants manning the phones. One is described as working the floor (of clients) 'faster than a stock market trader' through speed-reading the Koran (p. 58). In addition there are morality lessons provided by 'celebrity clerics', one owning a white Mercedes (p. 207). In complete contrast, mullahs are involved in prostitution services where the buying and selling of sex is legitimized as short-term marriage and divorce, payment for the latter constituting the fee.

There is a huge cosmetic surgery industry where nose jobs appear to be almost compulsory for the younger generation of both sexes. Many other parts of their bodies are similarly sculptured. This is operating in a world where 'sex is an act of rebellion in Tehran'. We are told that 'only in sex do many of the younger generation feel truly free'. It is a 'form of protest' with the usual 'deadly risks' involved (p. 179). And medics are on hand to rectify the consequences: 'Dr Sew-up' reconnects hymens in preparation for marriage (p. 263).

Tehran is also distinctive in an agglomeration that is missing: academics and intellectuals are conspicuous by their absence following 'one of the world's most spectacular brain drains' (p. 261).

So we can clearly see the city's uniqueness resulting from its relation to the state. But does this mean it is 'a city whose real life force is so suppressed under Islamic rule' (p. 4) as Navai initially claims? Later she finds 'the soul of the city' (p. 262) in downtown and describes pragmatic citizens eschewing the opportunity of moving up the Val Asr in favour of keeping to the freedom of the streets, something missing in north Tehran (p. 136). Such sentiments are typical of many cities. And it is diffusing its pathology of lying to smaller towns and villages beyond its boundaries (p. xiii–xiv). Tehran is a wonderful, vibrant city and Navai concludes by declaring her love for it (p. 281).

City Insights B: Timothy Brook's Vermeer, Delft and a Global World

In the 1660s and 1670s Delft was a small city in a small country – the Dutch Republic little more than half a century old – and yet Timothy Brook, an historian of China, claims that: 'Delft was not alone. It existed within a world that extended outward to the entire globe.' This is the key message of his book *Vermeer's Hat: The Seventeenth Century and the Dawn of the Global Age* (2008, p. 10). Delft was the home of the great Dutch artist Johannes Vermeer and Brook analyses five of his pictures (plus an image on a porcelain dish and a picture by another Delft artist) as a means to understand the making of a new global world.

How is this possible when 'Vermeer didn't paint anyone who wasn't born within 25 kilometres of Delft' (p. 210)? Brook's methodology is to focus on the props Vermeer uses in his paintings, objects that most certainly did come from far further afield. He uses them as 'doors to open' (p. 9) and view a wider world; he tells stories of life, death and work that underpin Vermeer's bourgeois life in Delft. For instance, the hat in the book's title appears in the painting *Officer and Laughing Girl* which stimulates stories of buying beaver pelts from Canadian Indian tribes to provide the quality felt needed for this newly fashionable attire (p. 43). But by following the money, Brook finds that the profits were used for further (failed) exploration to find a 'middle route' to China down the St Lawrence River and through the Great Lakes (p. 46). The general point is that 'The quest to get to China was a relentless force' shaping the seventeenth century, which is why 'China lurks behind every story in this book' (p. 19). These stories derive from props/objects that indicate trade in silver, tobacco, people and porcelain. Thus, although a specialist in Chinese history may not, at first, be an obvious author for a book on work in Delft (p. 230), actually he is ideally positioned for understanding this emerging global world.

And Delft is an appropriate starting locale because it had a chamber of the Dutch East India Company (one of six such Dutch cities), a major global player: Brook begins his book with Vermeer's *View of Delft* in which he points out the roof of Company's Building in the commercial heart of the city (p. 15).

Brook describes this new global world through its interconnectedness using the Buddhist mythology of Indra's net (p. 123):

> ... growing larger all the time ... As the density of strands increased, the web became ever more extended, and more tangled and complex, yet ever more connective. There were many spinners on this web, and many centers, and the web they made did not extend symmetrically to all places. Some places were favored more than others because of where they were and what was made there or brought to them.

And this involved an 'extraordinary cross section of humanity' evidenced by the passenger list of one Portuguese ship where the 'home' sailors were in a minority alongside Japanese, other Europeans, Muslim Indians, Filipinos, Arabs, Africans and Jews (p. 94). Brook describes all these stories as revolving around trade, ordinary people and states (p. 222) but this hardly does justice to his net analogy of nodes and links that are cities and the routes between them. Here is the A to Z of Brook's cast of the former: Acapulco, Amsterdam, Antwerp, Arica, Bantam, Batavia (Jakarta), Beijing, Buenos Aires, Canton, Delft, Fuan, Goa, Hague, Jingdezhen, Lisbon, Macao, Manila, Paris, Potosi, Quebec, Rotterdam, Seville, Shanghai, Suzhou, Venice and Zhangzhou. These are connected by myriad links with Macao–Manila, Acapulco–Manila, Amsterdam–Batavia and Lisbon–Macao being key trade routes (pp. 87–8). But I will focus on one of the other 'threads' in the net, movement of words and ideas (p. 124): the link between Jingdezhen and Delft.

Jingdezhen was China's leading pottery maker and exporter and developed the blue and white porcelain for which it was famous in responding to preferences in Persian markets centuries earlier (pp. 61–2). This high-quality product reached Europe in the seventeenth century as a luxury item. An affordable substitute was produced in Delft, not primarily as an import replacement, but for export. But importantly, 'Delft potters did not just imitate; they also innovated' by producing blue and white wall-tiles for the new houses of Delft bourgeoisie (p. 78). The outcome was an economic boom involving a quarter of the city's labour force in the new industry (p. 78). And thus Delftware – Chinese style, Dutch production – was born.

Pottery joined the Delft's school of painting as two remarkable city agglomerations in this one Dutch city that had worldwide impacts. However, perhaps this very successful specialization, like in its Chinese twin Jingdezhen, counted against further urban growth: both cities today are relatively small.

Finally, we should bring states into the argument as Brook insists. Cities are always vulnerable to state policies that destroy commercial activities. And so it was with the Tokugawa regime's autarkic policies for Japan leading to the demise of Macao, which in turn hit Manila, suffering also from China and Spain curtailing trade (p. 178). And to this we can add Vermeer's untimely death in 1675 at the age of 43. His livelihood required customers for his painting, which in turn required a prosperous local economy. In 1672 France invaded the Dutch Republic, purchases and commissions ended, and Vermeer was effectively bankrupted. Brook surmises that 'what killed Vermeer may well have been the same thing that gave him his career in the first place: Delft's place in the economic networks that stretched around the world' (p. 230). When the latter collapsed, so did Delft's famous painting agglomeration.

2. Cities as the birth of civilizations

Introduction: multiple civilizations

As previously reported, cities and civilizations are closely related concepts, the former being used to define the latter. This relation is most obvious at the birth of civilizations with cities as creators of more complex ways of living covering relatively large areas. In the original nineteenth-century telling, this instigating process was deemed to be a relatively rare occurrence, a few 'prime' civilizations from whence all others were descended. Such a definitive view of the development of cities and civilizations has not survived subsequent researches. New cities have been discovered in new regions worldwide and this has contributed to the erosion of the idea of civilization as a very exclusive club. There have been debates about whether discovered cities are the result of diffusion of urban practices or are themselves novel, latter-day city inventions, but this is not a major issue for this text. I proceed on the assumption of multiple developments of cities in networks across all major settled continents, culminating in the global-scale city networks of 'modern civilization'.

In this world of numerous urbanizations I focus on two contrasting examples. First, I describe the ancient civilization of Mesopotamia, a region that has a prodigious research tradition going back over a century of excavations and analyses. This is where cities were discovered to be creators of civilization as rich and complex societies. Hence we know more about the role of cities in the 'birth of civilization' here than anywhere else. Second, I explore the most recent discovery of cities in a civilization that has yet to be fully recognized. New archaeological techniques of infrared imagery from aircraft surveys have revealed the existence of a pre-Columbian Amazonian civilization through discovery of large cities along the rivers. Self-evidently we known much less of this urbanization, seemingly wiped out by diseases ahead of European penetration of the Americas, but what is emerging is an exciting revision of many long-held ideas on civilizations.

For both civilizations I use the five defining features of cities described in the previous chapter: process, agglomeration, connections, projecting power, and relations with states. Clearly, a focus on two urbanizations so very different – hugely separated in time and space – constitutes a good test for the transferability of the key city features and thus of this generic approach.

Sumerian cities and states

The earliest cities of Mesopotamia are found in the far south where the two rivers, Tigris and Euphrates, flow into the Persian Gulf. This is Sumer and cities emerged here in the 4th millennium BC and flowered in the 3rd millennium. To provide an initial sense of the size of this civilization I have used the population estimates of George Modelski's (2003) demographic inventory of cities over five millennia. For the earliest centuries he lists cities with estimated populations of 10,000 and above and some of his results are shown in Table 2.1. This features the five cities that reached a population of 40,000 or more by 2100 BC. Between them they provide a skeletal indication of the overall process of change in this urban world, a continual work in progress. The first great city was Uruk, the most southerly city, reaching 40,000 in the 4th millennium and peaking at 80,000 in the early 3rd millennium. This is the world's first great metropolis. And, of course, it was not alone. When we add the other cities in 2800 BC we find over a quarter of a million people who needed feeding each day, needed materials for their other consumptions each day, needed materials for the work they did each day, plus associated infrastructures of organization and transport to facilitate this production and consumption. Quite simply, this is a massive collective enterprise. In the subsequent urban developments, there is a gradual decline of Uruk to be replaced by competition between Girsu, Lagash and Umma, mainly related to the changing course of the Euphrates, before Ur becomes the world's first 100,000 city, Sumer's second great metropolis and capital of a huge empire stretching far beyond Sumer. This is the political outcome of early Mesopotamian work in progress; the consequent urban outcome is a network of fewer Sumerian cities, but larger ones.

Table 2.1 Sumerian cities

	3300 BC	2800 BC	2500 BC	2100 BC
Girsu	-	-	-	80,000
Lagash	-	-	60,000	-
Umma	-	20,000	40,000	20,000
Ur	-	12,000	10,000	100,000
Uruk	40,000	80,000	40,000	30,000
Other cities	20,000	156,000	140,000	70,000
Total population	60,000	268,000	290,000	300,000
Number of cities	3	11	13	9

The material basis of this new urban world derived from two sources. First, lands surrounding the cities were transformed into agricultural hinterlands based upon irrigation. Second, the vast dendritic river system was harnessed for mercantile activities, movements of commodities and ideas between cities and beyond. It is these environmental advantages that facilitated urbanization where both agglomeration and network externalities could create a great civilization.

Guillermo Algaze (2005a) defines the 'Sumerian takeoff' as a process of cumulative growth processes. He describes the growth of a textile industry in the 4th millennium involving the development of multiple specialist occupations from tending sheep to shearing, combing, spinning, weaving, plus an infrastructure of recording, storing, distributing the product while providing housing and subsistence rations for workers. The key innovation in the development of this infrastructure was the elaboration of accountancy practices based upon tokens, tablets and seals. A cuneiform script (markings on clay tablets) evolved as a bookkeeping tool thereby allowing a basic control over economic activity. It is these administrative advances that led to the cultural revolution that Sumerian civilization is most famous for: the invention of writing. Converting language into meaningful scripts on tablets produced a knowledge industry based upon a new profession: scribes. Together with training programmes, this became a huge industry producing the first large literature, stories in the form of myths and epics as well as more practical writings. The latter includes a 'Titles and Professions' list that indicates the complex division of labour – administrators, lawyers, officials, priests, jewellers, potters, bakers, coppersmiths, etc. - of Sumerian city agglomerations.

These city agglomerations required inputs from outside Sumer. For instance, new metal-processing industries were developed that needed imported ore and semi-processed ingots of smelted copper from mountains in Iran and Anatolia (today's Turkey). This formed the basis of what Algaze (2005b) calls the 'Uruk world-system' centred on Sumer as its economic core zone. Beyond this core a dependent periphery was constructed of dense trading links coordinated through trading posts and enclaves of Uruk's merchants. These took full advantage of Mesopotamia's dendritic transport opportunities to trade northwards to upper Mesopotamia and further into surrounding highland sources for raw materials. The key point of this economic process is that it was Uruk merchants seeking out trading partners to the north. They introduced a new large and expanding urban demand in Sumer for commodities to non-Sumer suppliers. Offering entry into this market proved irresistible but the Ukuk merchants had full control of transactions. It was an unequal economic exchange, a projection of the power of Sumer's dynamic cities.

Being at the very birth of a civilization, Sumerian cities' relation to state-building is novel. The development of cities as commercial hubs preceded the emergence of states as top-down political mechanisms providing social order, internal security and defence from outside. The need for the former resulted from cities becoming larger and thereby more multicultural. Bringing together diverse peoples into dense proximity inevitably generated social conflict, and city institutions changed to cope. Specifically, the organizational power of the administrator sector was superseded by a new military power of kingship. Prosperous cities were also targets for outside raiders; the king became protector. In these ways, Sumerian cities invented the state, Sumer's second major urban innovation, initially in the form of city-states. The conversion from commercial city to city-state is indicated by the building of city walls. And for Uruk, built about 3000 BC, they were huge: 9 kilometres long and 7 metres high, encompassing 900 towers. The labour required to build this necessary structure indicates the power of the king: commercial economic prowess is being replaced by military state power.

The invention of the city-state fundamentally changed inter-city relations in Sumer. Whereas commercial cities formed a network of cities based largely on complementary relations – partners in a league of cities – cities as states became rivals leading to military conflicts – winners and losers – resulting in more hierarchical city relations. The conflict we know most about was the rise of Umma on a new course of the Euphrates creating a long-term conflict with the previously more established cities of Girsu and Lagash. Nissen (1988, p. 35) dates this period of rivalry between city-states from 2800 BC to 2350 BC, after which two winning cities, first Akkad and then Ur, each established a city-empire by subjugating all rival Sumerian cities. Such multi-city states created new political problems; defeated cities had their walls destroyed, thus removing their statehood, but still had to be governed. This led to the invention of provinces, a territorial political innovation in which former city-states were subject to a distant centralizing power, a bureaucracy providing governors to rule on behalf of the imperial city.

The first such imperial city was Akkad, ruled by Sargon the Great, the earliest military leader to forge a great empire. He conquered all Sumerian cities as well as the rest of Mesopotamia and lands beyond. For 200 years his new city of Akkad, just to the north of Sumer, became the prime communication and administrative centre for the whole region. However, without a sizeable commercial base like Uruk, it never attracted a large population: Modelski (2003) estimates its peak at just 30,000 in 2200 BC. The actual site of Akkad has not yet been discovered. A short interval after Akkad's demise, the city of Ur performed a similar role in an empire that lasted a little over a century. However, with its new political role grafted on to a long commercial legacy, Ur did overtake Uruk for size as mentioned previously. However, note that both of the military imperial constructions of Akkad and Ur were fundamentally different from the previous dominance of Uruk. The latter never involved military conquest, it was an expansion based upon commercial enterprise. These two distinct processes, generating great cities as commercial centres and imperial capitals, have been interlaced in the history of urbanization ever since.

This earliest example of large-scale urbanization can be used to envisage how cities first emerged in other regions across the world. However, it should not be read as a simple evolutionary progression. Akkad plays the informative role here as interloper, taking advantage of Sumerian social development from outside. But the key point is always to separate the economic and political inputs so as not to confuse city with city-state. The invention of the state as a coercive process, whilst being a consequence of urbanization, had profound implications for cities. Henceforth, commercial practices in cities have had to develop alongside the increasing political power of states. Relations between cities and states have been an abiding feature of all future world urbanization.

Discovering hidden urbanization: pre-Columbian Amazonia

The story of Sumer's cities creating Mesopotamian civilization is a recounting of a long-term research effort – excavating sites and deciphering finds – in which urbanization is at the heart of the enterprise. Cities, some long known from the Bible, have been apparent in the terrain as 'tells', mounds built upon multiple layers of settlement. In complete contrast, recent researches showing the possibility of urbanization in ancient Amazonia have not been in such plain sight. Cities are hidden in two ways. First, there is the environmental context: unlike Mesopotamia with stone structures and metal artefacts, remnants of an Amazonian civilization built upon wood structures have long been consumed by the hot, wet ecology. Second, there is the cultural context: Amazonia has been habitually interpreted in modern thinking as pristine rainforest with the few human inhabitants blending into the ecology, making little or no impact. Recent advances in archaeological methods have changed both impediments. From above, viewed from the air, Amazonia's past urbanization has now been clearly sighted.

The form of archaeological investigation in Amazonia is ordered in a different way from traditional research. The latter is typically focused on specific sites that are then fitted into understanding a wider landscape, whereas advances by archaeologists in Amazonia begin by trying to make sense of large patterns in the landscape. These patterns, initially revealed through aerial photography and now augmented by in-depth remote sensing technology, are truly remarkable. Clement and his colleagues (2015) report on discoveries of hundreds of kilometres of causeways, tens of thousands of raised fields, multiple features of river management, plus numerous settlement networks of mounds, ditches, walls and roads. All these landscape constructions imply massive labour inputs indicating the organization of large populations.

This evidence is supplemented on the ground by the existence of Amazon dark earth, layers of immensely fertile soils from half a metre to two metres thick. These are the result of waste management around settlements and through agricultural practices (e.g. mulching) over several millennia. There has not been a comprehensive survey of these soils but they are thought to cover over 0.1% of the Amazon basin suggesting hundreds of thousands of hectares. Ceramics – pieces of broken pottery to be precise – are often associated with these soils. Without stone structures or metal tools these bits of pottery are the only industrial activity for which we have large-scale evidence. And it appears to be truly massive: Mann (2011) presents an estimate of many millions of pieces in just a single mound. Again this points towards large populations.

Large populations need to be fed: what has been grown in these very productive dark soils? Here archaeologists have drawn on the work of crop geneticists. The latter indicate that Amazonia is an important world centre of plant domestication; among the hundred or so species domesticated here are manioc, sweet potato, tobacco and pineapple. Over half of the domesticated species are trees – fruits, nuts and palms. Clearly the Amazonian rainforest has been truly bountiful, well equipped to support a large population.

So how are we to interpret this emerging knowledge of a region traditionally viewed as sparsely populated? Unfortunately there is no evidence of writing so the voice of these ancient peoples is not available. We can never reconstitute their story in the way created for Mesopotamia. However, there has been renewed interest in the earliest European report from an expedition that travelled by boat from the Andes through to the mouth of the Amazon in the 1520s. Previously dismissed as more imaginary than factual, there are descriptions of numerous very large cities along the banks of the Amazon and its tributaries, reports that are now being reassessed. These cities would have provided the demand that led to the invention of production of new foods (plant domestication) and which would have diffused as innovation along the rivers as communication corridors. This is a civilization lost to history through the decimation of indigenous Indian populations by European diseases in the first decades of transatlantic contact. The resulting depopulation and demise of cities left the Amazon forest obviously not 'pristine' as later interpreted, but rather as an overgrown, obliterated ancient urbanization. Thus, today's empty patches on archaeological maps where nothing significant appears reflect the fact that only a small portion of Amazonia has been studied: these should no longer be assumed as being historically 'empty', rather they represent opportunities for new research, perhaps revealing more cities.

Concluding supplement: thinking beyond past urbanizations

Of course the link between large rivers and the emergence of city networks is a well-worn path: until the recent past, movement over water was hugely more efficient than over land. In taking advantage of such enhanced trade and communication potentials Amazonian urbanization joins not only Mesopotamia but also replicates other major ancient civilizations along the Nile, Indus and Chinese rivers. To these we can add more recent work on urbanized landscapes, for instance centred on the Mekong, upper Niger and pre-Columbian Mississippi rivers. But why should we be interested in these cities of long ago, surely we have enough contemporary urbanization – twenty-first century as first era with a majority of people living in cities – to study and understand?

This is a very reasonable query given our previous description of knowledge overload as a problem in the Preamble. However, there are several reasons to include earlier urbanizations in this advanced introduction. First, there is the important matter of curiosity. We know cities because we live in them and/or visit them, but such familiarity is inevitably limited. It has spiked an interest in urbanization – why else would you be reading this book – but however many cities are visited they will inevitably represent a biased sample, most likely geographically, most certainly historically. We must understand beyond our experience; this is to be curious, the essential beginning of any study and understanding. Second, there is a need for broadness, not completeness, in trying to make sense of cities. This breadth is necessary in order to appreciate the centrality of cities to the human condition.

It is the very explicit integration of ancient cities with past civilizations that provides a constant reminder of the crucial importance of urbanization. Continuing this thinking into the present brings us to the question of the role of cities in modern civilization, a way of living that has created the global climate emergency. This will be the subject of the final chapter, where the idea of modern cities as the end of civilization is broached. Between these two forays into matters of civilization, we build up our knowledge of cities per se, in multiple ways both past and present. As a relatively recent invention in the evolution of our species – somewhere between the last 2% to 5% of *Homo sapiens* existence – the pervasiveness of cities across the world really is astounding.

City Insights C: Brenna Hassett's Bioarchaeology

According to Brenna Hassett (2017) in her book *Built on Bones: 15,000 Years of Urban Life and Death*, 'we haven't just built cities, cities have built us' (p. 14). This remarkable statement is from a bioarchaeologist who uses the latest scientific methods to glean information on the past from the skeletons retrieved from excavation sites. Although not from an urban studies background she shows in her writings that she really does get cities. She asks 'If cities are so great, why are they so full of things that kill us?' (p. 9) and thereby highlights what is one of the most intriguing things about the incessant rise of urbanization. For nearly all of their existence, death rates in cities far exceeded birth rates so that growth occurs because many people are attracted to cities despite the obvious mortality threat.

Settling down in fixed locations resulted in physical changes to the human body. Compared to hunter-gatherers, urban populations had 'smaller frames, weedier legs and shrinking faces' with resulting dental problems (p. 79). And there was a baby boom: gaps between births were reduced with the old problem of babies inhibiting movement as a way of life removed. But because cities became 'mortality sinks' (p. 306), this 'kept a lid on the fertility explosion' (p. 29). Cities altered our health forever through 'changing the environmental reservoir for diseases' through 'increasing connections' (p. 206). We became 'meals on wheels' for bacterium (p. 225).

The key process is the conversion of scattered rural infections into concentrated urban disease so that 'the real contribution of the invention of the city to the epidemiological transition is the shift from infection to *epidemic*' (p. 211, italics in original). Hassett describes the process thus:

> most of our modern epidemic diseases are actually viral infections direct from the rustic goodness of the soil. They would stay there too, but in a networked world of roads, buses, cars and people, they now reach the dense urban populations they need to explode. It's the ready transmission of infection that makes a disease into a plague; the networks of roads, trade and people who link an increasingly urban world. (p. 211)

The tracing of tuberculosis through changes in skeletons is discussed, showing clear increases where cities are thriving and prosperous. For instance, there was a tuberculosis 'uptick' in Roman cities of the early AD centuries: 'not an unexpected finding given the well networked and urban nature of the Roman Empire' (p. 205). A little later the Justinianic Plague arrived from the East via the Silk Road (p. 220).

But Hassett argues that it is in the fourteenth century that 'the transmission of infective diseases ramped up to the point where virulence outbid its rivals in disease forms' (p. 205). Welcome to pandemics:

> Epidemics of infectious disease became *endemics*, naturalised citizens of cities thanks to the contributing factors of population growth, mobility and integrated networks for the transmission of people, goods and animals, all driven by the engines of urban growth. (p. 205, italics in original)

Thus did the Black Death arrive from Caffa into Genoa in 1348 to spread across Europe and linger for several centuries. And there is also the (possible) swapping of diseases with the Americas coming into Europe's orbit: smallpox decimated indigenous Americans, who gave Europeans syphilis in return. The latter spread quickly through Europe's cities: 'The nature of syphilis is relentlessly urban' (p. 250). And, of course, this is all a consequence of 'the power of cities to draw in labour, and to exploit it' (p. 271).

Beyond these 'natural' disasters, Hassett powerfully portrays the relations between cities and states:

> Cities are the harbingers of complex polities; they are defined by their complexity and are the necessary engines of the state systems that contest territory and power. (p. 186)

Violent deaths can be discerned from skeletons and the nature of the homicides change from early 'raiding', often killing across a victim community, to the more organized warfare that comes with cities and states where the organization of young males as warriors creates age and gender specificity to finds of multiple skeletons.

My description of Hassett's text is perforce limited to selected highlights and as such misses much of the nuance and careful discussion: she is particularly good when covering debates on the meanings of various findings. And her subtle scholarship transfers to her understanding of cities. She is reticent to simply define what a city is with all the loss of flexibility that that would entail (pp. 94, 115) and concludes by admitting she has 'skated around defining a city' to focus on 'features ... that changed us'. I interpret her position as treating cities as process.

Finally, given that Hassett's book appeared in 2017 and I am writing this summary in 2020, the question arises as to whether she had the prescience to see the Covid-19 pandemic of the new decade. She does refer to 'the globalisation of the urban world … by cities in contact with other cities' and goes on to say 'it's not hard to imagine that our globalised network of disease could come back to haunt us some day' (p. 305). After all 'cities attract people, and people carry diseases' (p. 190), and 'the urban tentacles, eventually, get everywhere' (p. 307).

3. Busy cities

Introduction: layers of busy-ness

Cities are first and foremost very busy places. This is exemplified during 'rush hour', the period of the morning and evening commutes when traffic peaks. Rush hours are ubiquitous across the world; when lists of worse congestion are published, cities such as Los Angeles, Bangkok, Istanbul, Mexico City, Moscow, Jakarta, London and Guangzhou are typically featured alongside pictures of polluting traffic jams. Estimates are made of the many hours per week typical commuters allocate of their precious time to this unhealthy, stressful activity. It is a classic example of a negative feature of urban living that has to be endured to obtain the positive advantages of working in a city. In other words you cannot wish it away – a diminution of traffic would signal a city in decline, not something you want to add to your urban experience. The formula is simple: the greater the local rush hour, the more economic activity providing local job opportunities. And this works for all cities not just the major cities mentioned above.

The simple point is that each individual trip that constitutes rush hour, by car or public transport, has an origin and a destination; if these travel lines are mapped they depict a massive splotch of criss-cross cutting links. Seemingly unfathomable, this represents the complexity of the city in its most explicit guise. Yes, busy, busy, busy, but not unfathomable. This is merely the framing of the busy-ness of cities, its scheduling on a daily basis. The defining 'busy-ness' of cities is the work done between the two daily commutes. This is the business of cities, its amalgam of myriad economic activities. They are the product of urban growth, the dynamics of cities ever changing through developing new work whilst discarding old work.

It follows that at any one point in time a city is composed of historical layers of economic activities that represent past product cycles, from development to demise, within its different industries. Today the generation of the latest work is typically signposted by a skyline of huge cranes involved in building new offices. But the commuters are largely going to the established work of previous but recent economic development, others to long-established work, some of which is beginning to decline. In this way a city economy consists of business 'strata' created by cycles of past 'busy-ness'. Some discarded work may even be recycled as heritage to provide new tourism work. And all this fundamental dynamism stems from agglomeration processes.

As described in Chapter 1, agglomeration processes are most commonly viewed as economic processes, the benefits they provide are treated as externalities. These city advantages are outside normal market practices in provision of products and services. Thus most of this chapter will explore the operation of agglomeration externalities. But the effects of dense city populations, individuals in everyday interaction with each other, have ramifications beyond production for markets. Most obviously cities are places of ultimate consumption of commodities – shopping! Shopping centres in their various forms are the most conspicuous agglomeration; they depend on being busy. But beyond the sphere of economics there are inevitably other agglomeration effects - finding new ways of doing things – within social, political and cultural behaviours. These will be briefly introduced below, and will be developed further in later chapters.

City agglomeration and sectorial clusters

Agglomeration externalities have been theorized in two quite different ways. Economists, starting with Marshall's 'industrial districts', have tended to focus on clusters of producers creating similar commodities. The idea is that interactions within dense spatial clustering of like-minded entrepreneurs, typically small and medium-sized firms, will create a successful economic locale, capable of reproducing itself in innovative, new and productive ways. Put simply, producers learn from each other, seeing a neighbour improve a production process soon results in copying across the sector. Thus clusters of individual firms combine market competition with non-market complementarities. Historically, the clustering of industries in different quarters of cities is commonplace; in the twentieth century such concentrations were central in developing key elements of what we understand as 'modern' society. For instance, New York's Madison Avenue became synonymous with advertising; similarly, London's Fleet Street meant national newspapers. And, of course, both have their designated financial clusters in Wall Street and 'The City' respectively. These examples join with Los Angeles' Hollywood, introduced in Chapter 1, as world-changing economic clusters. In all cases it is postulated that producers within the cluster have market advantages over those producing the same commodity outside the cluster. Thus the latter producers will suffer relative market decline while those in the cluster prosper, thereby enhancing the economic clustering.

The second approach treats the agglomeration externality as a consequence of the city considered as a whole. In this way of thinking there are advantages to be found in the immense variety of economic activities that constitute a city. In particular, this enables a much wider learning process across different sectors. This process produces 'Jacobs externalities', named after Jane Jacobs, the famous urban activist and scholar, who developed a theory of economic development through cities. Her work does not preclude the importance of clusters – the learning can be inter-cluster. For instance, Los Angeles' city-region economy consists of much more than Hollywood, with clusters of jewellery, clothing, furniture production, automobile design plus new tech hubs, all creative industries brimming with potential innovation links (Scott 2008). But Jacobs warns about the over-success of any one cluster. For her, economic diversity is fundamentally important and this is threatened when a single industrial cluster grows to dominate a city. Without a broad range of successful industries, the city's future becomes hooked into a single product cycle with inevitable consequences on the downside. Detroit has become the prime example of this process; with the waning of its automotive agglomeration externality its urban decline was inevitable.

Edward Glaeser and his colleagues (1992) carried out an important study that attempted to assess the relative merits of these two approaches – specialist clusters and citywide agglomeration. In a large-scale analysis of 170 US metropolitan areas from 1956 to 1987 they focused on the six largest industries within each urban area. They measured how employment in these city-industries changed over the 30-year period. The employment change was compared to measures of economic specialization and economic diversity. With a total of over 1,000 city-industry observations they were able to show that with more concentrated industrial structures, implying specialized clustering, employment growth actually slowed, the opposite of what is expected of clusters. In contrast, the more diverse the industrial structure the greater the employment growth, in line with Jacobs' expectations. The authors insist these results be viewed tentatively. For instance, they point out that the specific period covered was dominated overall by lower economic growth and therefore perhaps was not conducive to clustering mechanisms. But the very strong empirical support for citywide diversity is an important finding.

Of course, there is another reason we should be tentative in interpretation of the finding on clustering: industrial clusters come in many different forms, some of which may not be very conducive to generating economic externalities. For instance, there are different types of clusters in terms of their spatial form and inter-firm relations (Markusen 1996). In addition to Marshall's industrial districts – multiple firms in a single sector – three other cluster arrangements have been identified: a hub and spoke form centred on a few dominant firms surrounded by suppliers; satellite platforms as congeries of branch factories taking advantage of particular place attributes such as low wages or state incentives; and state-anchored clusters with firms taking advantage of supply opportunities afforded by large public institutions. In each of these other cluster types small and medium-sized firms are dependent on larger entities, and this relatively weak market position is likely to stymie economic externalities. But in the large-scale analysis reported above all forms of economic clustering are lumped together in the measurement of industrial concentration. Thus the economic externalities that Marshall first identified cannot be dismissed. But what this analysis does suggest is that such potent economic clusters are perhaps not ubiquitous and are best identified through case studies.

It is also the case, of course, that cities come in many different spatial forms that might impinge on the operation of agglomeration effects. The metropolitan areas that are used in the Glaeser study are defined by the US census and are based upon a model of a central city surrounded by residential suburbs. While this urban form was common in the 1950s, subsequent decentralizations of many economic sectors have produced increasingly complex spatial structures. This has resulted in cities becoming larger, defined as 'city-regions', many still central city-centred such as in London with southeast England, but also often coalescing to produce 'multi-nodal city-regions'. These can be very large, based upon the first so identified as 'Megalopolis', the east-coast US region from Boston to Washington (BOSWASH). These multi-nodal city-regions can be important in extending agglomeration effects. This is referred to as 'borrowed size', where a smaller city can take advantage of the economy of a nearby larger city and its wider range of services. Thus, for instance, in the city-region of Randstad Holland (Amsterdam, Rotterdam, the Hague, Leiden, Utrecht) the creative industry of Utrecht can take advantage of the more advanced financial services in nearby Amsterdam for its special capital needs, an economic advantage that would not normally be available to firms in a small city. And all of the Randstad cities can take advantage of Amsterdam's Schiphol Airport providing physical connectivity at a global scale. But borrowed size in not always a positive feature: Silicon Valley in the San Francisco Bay Area multi-nodal region has exported its housing demand to San Francisco city itself with serious negative economic effects on the central city. We consider multi-nodal regions further in Chapter 8.

In many ways the high-tech cluster of firms that has developed over the last half-century in the Santa Clara Valley to the south of San Francisco is a classic case of a Marshall industrial district. It has been a process of multiple firms combining market competition with cooperative development creating immense innovative success. It is the most successful new clustering process and this has led to many attempts at imitation globally. However, the key point is that it is a process, an unfolding of a set of practices over many decades. As such it cannot be simply transplanted; its uniqueness has to be recognized. Lessons can be learned, notably the importance of fostering cooperation in a fiercely competitive sector. But other high-tech clusters across the world will necessarily be very different: for instance, the successful IT hub in India's Bangladesh features strong national and provincial government support. This discussion is a clear example of the generic/specific framework introduced in Chapter 1. Whereas the clustering process itself can be viewed as a generic development, specific amalgams of practices are not transferable.

This conclusion is very important. By their very development, local sectorial clusters are economic success stories and therefore they attract the attention of urban policymakers. But if replication is problematic it follows that economic externality returns will be rare. In contrast, the

Jacobs externalities based upon citywide diversity are readily available in all cities. Busy cities are intense concentrations of large populations with multiple backgrounds interacting through their everyday behaviours. This maelstrom is a breeding ground for invention and innovation, the job of policymakers is to harness it in such a way as to generate positive economic externalities. If successful, this will likely lead to home-grown sectorial clusters: a local unfolding of the process signifying new economic growth.

Eclectic agglomeration effects

Cities are inherently diverse and therefore we should expect multiple effects of agglomeration beyond economic production of goods and services. There is no systematic literature that explains these other agglomeration effects; here they are simply listed and briefly described with their associated innovations.

- *Retail sector*. This sector originates in local food markets and trade fairs. It was transformed in the nineteenth century as rapid urbanization and industrialization transformed the size of cities and thus their consumer market. Innovations in shopping included large departmental stores in the second half of the nineteenth century, chain stores in the first half of the twentieth century, increasingly large supermarkets in mid-century, and shopping malls in the second half of the twentieth century. The result has been that all cities are popularly recognized for their retail offerings.
- *Entertainment industry*. There is a similar sequence of innovations in this sector, from the growth of a range of theatres in the nineteenth century through cinemas in the twentieth century to large arenas today. Traditionally they co-located with retail, especially restaurants, to create 'downtown', the exciting centre of cities.
- *Popular sports*. These are specific innovations of the late nineteenth century when selected pastimes were codified and organized by clubs attracting large city fan bases. Association football in Britain and American baseball are the classic cases, creating new popular markets and thereby creating the suite of activities that constitute the contemporary sports industry worldwide.

- *Public sector.* Although often derided as 'bureaucratic' – doing things by the book – this is a sector that has grown immensely in large clusters within cities thereby providing the conditions for innovation. And these have been forthcoming as workers find better ways of doing things. There is a different incentive; market advantage is replaced by political or career advancement.
- *Capital cities.* In the modern world there has been a practice of separating the capital city from the main economic centre; most large US cities are not state capitals, some countries have created brand new capital cities, one of the first being Washington DC. A motive for this has been to separate economic interests from political power (i.e. curb corruption) but this has crumbled with the rise of the economic lobbying industry whose members typically outnumber politicians in these cities.
- *Criminality.* Cities have traditionally been seen as sinful, and opportunities to behave outside the law also increased with the remarkable rapid urbanization from the nineteenth century. All cities developed criminal districts, both tolerated and condemned. Where else would you find, say, counterfeiting skills except in a city? (The earliest systematic study of cities focused on Chicago through the 1920s and 1930s where crime was integral to the studies.)
- *Political protest.* Cities are far too complex for governments to fully control and therefore are prime sites of resistance. Where these are tolerated it enables protest movements to use the city to develop their repertoire of resistance. With authoritarian governments cities become centres of subversive activities beyond the authority of the government.
- *War and foreign occupation.* Cities being besieged, or bombed, or under foreign military occupation have their everyday lives badly interrupted. Despite despair people find they are not powerless: they are still an agglomeration and they find ingenious ways of survival in terms of food, housing and healthcare.
- *Migrants.* From the second half of the twentieth century, cities in Africa, Asia and Latin America have grown immensely – some becoming megacities – despite being located in poorer countries. Despite these dire circumstances, new migrants have found ways of surviving by finding work and making work. Not usually recognized as such, this is agglomeration, providing opportunities and people innovating and imitating to take advantage. We return to this below.

This list provides a flavour of the great variety of city processes resulting from the power of their agglomerations. None of the examples is in any way exotic; they are all everyday practices to be found in different combinations across cities. This is an affirmation of the sheer complexity of cities. Subsequent chapters will elaborate on some of these themes.

Concluding supplement: megacities

Possibilities for generating agglomeration externalities are related to the size of cities: more people equal more interactions equal more opportunities. But it is not a simple causal relationship; externalities vary in quantity and kind depending upon many other characteristics of cities. This is important for understanding a well-known category of cities called megacities. These are defined by United Nations agencies as urban agglomerations with populations of over ten million. Because of varying population estimates the number of megacities is not agreed upon but there are definitely over 30 in the world today. Only seven are in rich G7 countries (London, New York, Los Angeles, Nagoya, Osaka, Paris and Tokyo), China has the most for one country (Beijing, Guangzhou, Chengdu, Jinan, Shanghai, Shenzhen, Tianjin and Xiamen), but most are located in poorer countries. And herein lies a problem: although agglomeration processes are generic they have produced very different outcomes across megacities.

All megacities have their high-rise skyline, housing a local cluster of a rich business class that is connected globally. But beyond this singular mechanism, other city processes, especially in poorer cities, generate a very different world. They have grown to megacity size by attracting large numbers of migrants who expect to improve their lives. The agglomeration of so many people provides opportunities not available from whence they came. The result is a plethora of innovations and imitations creating new work but in a very constrained context. This is city entrepreneurship but it is dominated by the basic goal of economic survival, quite unlike the economic growth found in richer cities. Thus it is not surprising that the policy issues raised in the megacities literature are all about urban problems, usually termed 'challenges', such as slum housing, crime and corruption, traffic congestion and air pollution. Of course the latter problems are not unknown in richer cities but they affect life chances to a far greater degree in poorer cities. And this is not merely a matter of classification, of corralling poor cities together; it is a functional outcome of the position of the cities and their countries in the world economy.

So, once again, population size is a very important variable for understanding cities since it causes the agglomeration effect but it is only a single measure. As such it should not be expected to characterize cities on its own. New York and Kinshasa are both megacities but treating them together has little utility for understanding either city. For this reason the concept of the megacity has limited utility and will not feature further in this book.

City Insights D: Luiz Eduardo Soares' Rio de Janeiro

According to Luiz Eduardo Soares (2016, p. 2) 'Cities are like snakes that shed their skin'. He was brought up in Rio de Janeiro, became an activist and politician, and is currently a humanities professor. His book, *Rio de Janeiro: Extreme City*, is a contemporary history of the city from the Brazilian military coup in 1964 to the great Rio demonstration of over a million people in 2013. It is a collection of stories, a mixture of biography – revealing stories as witness – and harrowing stories from interviews. The period covered starts with despair, moves into hope and ends in disarray – this is his skin metaphor quoted above. It relates directly to the notion of a city always being a work in progress. But incessant incremental change does not undermine structural confines of a city's development: once a snake always a snake?

What is this critical confinement in Rio de Janeiro's case? First and foremost, being a Brazilian city it has immense social inequalities. The most obvious representation of this is in the stark geographies of myriad cramped favelas for the poor and comfortable living and working places for the middle classes. And above both are the mega-rich and their operatives. Soares refers to the latter as the 'dynamic hub of metropolitan Rio' consisting of 'almost every mayor from the region … judges, chief justices, captains of industry, celebrities, footballers'. And you get the idea that agglomeration works differently in this city through inclusion of 'entrepreneurs who shoot their way into business' (p. 200). The external practices of this elite are built upon a 'network between London and the Amazon rainforest' (p. 264) and take in Brasilia (national politics), Miami (alternative drugs market) and Las Vegas (recreation) (p. 201). It is here that we find 'power, the power not to use power, and the simulation of power' that does not 'discriminate along party lines … we elect together, and we govern together' (p. 200). Chillingly, the claim is that 'there is no such thing as opposition' (p. 199).

Soares enters the fringes of this elite world with the rise of the Workers' Party (PT), becoming National Secretary for Public Security when PT's leader, Lula, becomes President in 2003. This followed a similar security role he held for the Rio de Janeiro state government in 1999. The author's experience in these security positions is central to his understanding of Rio. In a very illuminating episode, Soares joins Lulu in a listening exercise with people from the favelas invited to air their grievances. To

Lula's bemusement, one after another relays stories complaining about the police; why, he asks aren't they engaging with his basic social democratic programme of health, education and unemployment policies? Lula doesn't get Rio (pp. 3–5).

Rio's favelas population lives in territories marked out by rival drugs gangs who violently compete with each other within a corrupt and malevolent police overlordship. Soares argues that the police forces are ungovernable – 'not so much institutions as archipelagos of relatively autonomous units, each with its own remit and appetites' (p. 81). They kill with impunity. The base of their power is their role as a buffer between the 'dangerous' favelas and the everyday lives of the city's middle class. Therefore, they need to nurture the threat of danger to maintain their rule thus becoming forces of insecurity. This is overtly revealed in the 'policing' of the 2013 demonstration in Rio: 'the police force were bringing on exactly what it is supposed to combat. The guardians of public safety were taking delight in provoking chaos and even more delight in quashing it by force' (p. 256).

What is the common politics that sustains this inequality? Soares starts with pragmatism: police killings become 'routine' in their 'iniquity' (p. 94) so that 'in Rio you just get on with it' (p.103). There are complaints directed at an indefinite and vague 'they' but without a consequent 'us' (p. 250). The great 2013 demonstration confirmed the top-down social democracy of PT as part of 'they'. But Soares finds hope in the mass of contradictory demands of the participants: 'the propagation of a generous individuality could be the chance we need to reinvent collectivity in a more democratic and just light' (p. 253). This is a form of resistance he had earlier noted in reaction to a police killing: 'the pure power of collective intelligence' (p. 86). Alas, this hopeful political skin has lately been well shed.

Soares wrote his book to counter what he terms the cliché of Rio de Janeiro as a tourist destination: a fun city of beaches and bikinis. He does not deny this idealized image, just its ability to hide an unembellished complexity that is the Rio de Janeiro he knows and wants to tell the world about.

4. Cities connected

Introduction: cities come in groups

Cities grow through their agglomeration processes, but not by them alone. Agglomerations are promiscuous. By this I mean that the intense networking within cities operates in conjunction with further networking between cities. The individuals and firms that constitute an economic agglomeration through their work are commonly connected to similar agglomerations in other cities. These may be material links (consuming external commodities), personal links (meetings with like entrepreneurs), knowledge links (diffusion of innovations), or organizational links (operating through many sites, service links through such as multiple offices), all of which require physical infrastructures enabling contact between cities. These links are not optional add-ons to the work of agglomerations; they are necessary for maintaining the vibrancy of agglomerations across all cities. Thus do cities historically always come in groups; the lone city (usually conveniently 'lost' or an abstract myth) has never existed. Today many of these city connections are global and are increasingly electronic.

It is known that long-distance connection between human groups occurred for many millennia before the development of cities. This is indicated through the known geological origins of rocks used for making tools and ornaments. Two well-known examples are widespread finds of sharp cutting implements made from obsidian mined in just a few sites, and the extensive geographical distribution of amber beads exclusively from sources on the Baltic shores. These items passed along multiple agents in chains of connection and communication; as they intensified in number they focused on specific trading camps for both exchange and refining raw materials. It is here that we can find a growing community of specialists; when they increase to become embryonic agglomeration of productions the basic building blocks for city development come into play. The chains become more organized as the proto-agglomeration's needs become continuous to create a rudimentary network of cities, trading with each other. It is from such humble beginnings that the civilizations from Chapter 2 derive. Such networks help make cities but how are they themselves made?

Making city networks

Following Castells' (1996) concept of spaces of flows, there are two parts to this story: the infrastructure of movement – how this physically was achieved; and the simultaneous transfer of people and ideas – how distant humans interact. Although obviously closely intertwined, these two problems required very different types of solution to enable steady trading and thus cities.

Enabling long-distance movement of people and goods has involved a historical series of technological advances. The starting point for discussion is the point made in Chapter 2 that transport on water has historically been far more efficient that over land, hence the prevalence of early civilizations along major rivers. In fact most of the leading cities in the world today trace their origins back to early port functions. Thus historically, the main relevant technology changes have been in boat construction: river and coastal vessels to cross-sea and ocean-going ships, all by oars and sails, before modern steamships – dependable in all seasons – culminating in today's huge container vessels. On land the technology changes involve invention of the wheel and use of beasts of burden – donkey, camel, llama, etc. – travelling in groups ('caravans') along known roadways. Of course these two means of movement were interconnected – key roadways were mountain passes between river systems. However, it is in this latter form of transport that modern technologies have been most revolutionary: the worldwide spread of railways in the nineteenth century and the twentieth-century love affair with roads (bikes, cars and trucks) have consigned the horse to history, film and sport. And the modern package on movement has been completed with global air travel and global electronic communication. The end result is the concurrence of twenty-first-century people being by far both the most connected and the most urban in history. This is more than a mere correlation, there is a two-way causal link: this is Castells' 'network society' superseding modern industrial society, wherein he identifies 'global cities' replacing 'industrial cities'.

In terms of city networks these sequences of technological innovations have created inter-urban links beyond river systems: historically in seas, such as the South China Sea, the Black Sea, the Mediterranean Sea and the Baltic/North Sea, and larger afield such as the North Atlantic, the Indian Ocean plus, on land, the Silk Road linking East Asia with Western Asia and Europe. All these examples stimulated vibrant cities, which in turn assured a flourishing network. In the modern world this networking process has become worldwide, represented today by the world city network, to be described in Chapter 8. The effect of the modern technological advances has been accelerating space–time compression: travel times between cities have been radically reduced, for communication the reduction has effectively eliminated distance – satellites and cables enable instantaneous global connections. Thus are today's financial markets – historically a key city function – continuously open in all time zones across the world.

Although the initial modern 'erosion of distance' benefited and stimulated city growth by speeding up production and consumption, it has been suggested that the latest development, instantaneous communication, represents the 'end of geography', with the corollary that cities are no longer necessary. With office rents astronomically high in London, why not relocate and do all your financial transactions on, say, the Isle of Skye in northern Britain, not just for the much cheaper rents but, more generally, for a new, friendlier lifestyle? This is not happening. Why? We return to agglomerations. Number one amongst the requirements for successful economic transaction is trust. This has to be built and sustained through personal contact and communication. In the past it was achieved through traditions of hospitality for strangers and ultimately creating new inter-group languages, for instance Swahili for Arab–East African relations and various versions of 'pidgin English'. In the modern corporate world trust is achieved through face-to-face meetings, the coming together of people working in different cities. Success depends on a pooling of commercial knowledges gleaned from a variety of agglomeration experiences. This is why international airlines make so much of their profit from business-class customers, video conferencing notwithstanding.

Of course, people living in cities use them for much more than work: for our putative financial traders in Skye, recreational agglomerations in London would be sorely missed. It would be a brief, failed geographical experiment. So what is this contemporary world city network with its immense connectivity infrastructures? The key agglomeration is to be found in Saskia Sassen's (2001) process of global city formation. The initial key innovation comes in the 1970s with the integration of two erstwhile distinctive industries: computing and communication. Combined they enabled two related geographical developments: economic dispersion with concomitant concentration. The former resulted in the 'new international division of labour' whereby production was moved from traditional industrial zones to poorer countries to take advantage of cheap labour and less regulation. But this economic reorganization required new corporate governance structures and business servicing support that became strongly clustered in a few cities.

Sassen termed these 'global cities'; special places where new demand and supply agglomerations developed. These were where corporate headquarters coexisted alongside clusters of financial, professional and creative service firms that aided the corporations to negotiate complex transnational markets. As well as being international financial centres with myriad banking services from capital movements to wealth management, these cities housed law firms providing legal advice on multi-jurisdictional contracts, accountancy firms providing auditing checks across different countries, advertising agencies promoting products globally, and management consultancies aiding worldwide corporate strategies. This supply of advanced business services to satisfy demand from large corporate entities defines the emergence of global cities in the 1980s. Sassen focused on just three such cities – New York, London and Tokyo – and briefly mentions some others such as Paris and Hong Kong.

Since this initial identification of global cities, the business services themselves have become increasingly global. Traditionally servicing national markets, these firms at first moved to pastures new by following their clients to foreign locales in order not to lose them. But once relocated, it made sense to find new clients in the new market. Thus the service firms, first in finance and then across other services, themselves became global corporations. This process snowballed so that by the new century these firms typically operated through multiple cities across the world. In this way global city formation transmuted into a much broader world city network formation process incorporating hundreds of cities. These are recognizable by their offices in skyscraper agglomerations in the centres of large cities on all settled continents. Each company – in finance, accountancy, law, advertising and various consultancies – has its specific office network depending on its country of origin and expansion history, but in aggregation they constitute the world city network that enables economic globalization to operate and grow. Analysis of this network shows that it is dominated by London and New York, but there are several other prominent cities such as Paris, Hong Kong, Tokyo, Beijing and Shanghai and a geographically wide range of other globally important cities such as São Paulo, Mumbai, Moscow, Sydney and Seoul. But the key point is that this network process penetrates into the urban economies of hundreds of cities worldwide from Aberdeen to Zanzibar.

This global networking – a pooling of knowledges from multiple business service agglomerations – is not, of course, the only activity to take advantage of communication technologies to network globally through cities. Here are some other well-known examples.

- Energy industry networking featuring Houston, Calgary and Perth as leading cities.
- Non-governmental organization's work organized around Geneva, Nairobi and Bangkok.
- Film production focusing on Los Angeles, Mumbai and Hong Kong.
- Islamic financial servicing centred on London, Dubai and Kuala Lumpur.
- Diplomatic networks based in capital cities with Washington DC, Beijing and Moscow being particularly influential.

Each of these networks is much wider than their listed trios of principal cities above and they represent just a small sample of global working relations through cities. Overall they constitute an extremely complex inter-city network ensemble, within which the world city network created by advanced business services described previously may be viewed as a sort of core mainframe, which we call the world city network, and which we explore in more detail in Chapter 8.

Explosive city growth

In the previous chapter Jacobs' externalities were featured as an important economic advantage, key agglomeration effects, for doing business in cities. These derived from the diversity of activities, and thereby commercial opportunities, only available in cities. But cities are also diverse in their connections, as this chapter has shown. Jacobs (1970) has combined agglomeration and connectivity in her economic growth theory through cities. Her argument centres on the fact that city growth typically does not follow a smooth trajectory; rather, relatively short periods of intense economic spurts are found. This is at the heart of her understanding of cities: in fact she defines cities explicitly as settlements that have experienced at least one such economic spurt. For some cities there has been just one spurt. She uses the example of her home city, Scranton, Pennsylvania, which expanded rapidly only in the first few decades of the twentieth century. In contrast, large cities have experienced multiple spurts resulting in their immense economic diversity. Jacobs views these spurt episodes as unleashing immense economic power, hence her designation of them as explosive city growth.

Jacobs' description of this process is full of empirical examples, both historical and contemporary. One typical case is Los Angeles at the end of the Second World War. The city had been at the forefront of the war effort in the Pacific, but after 1945 its two leading industries, aircraft manufacture and shipbuilding, rapidly declined. However against all predictions Los Angeles thrived. A new economic spurt resulted in its economy soon becoming larger than it was before the war; by 1949 one-eighth of all new businesses in the USA were opening up in Los Angeles, covering a vast array of different productions. This city was leading America's post-war boom. We can add an even more remarkable turnaround of this period that was occurring in Germany, literally from the ashes of the war. Flattened cities were being rebuilt physically, and within a decade they were leading Europe's economic post-war boom. Both success stories were about creating new agglomeration effects in an increasingly connected world. But what was the actual mechanism of such explosive city growth?

The specific links between agglomeration and connectivity that generate economic spurts are import replacement with concomitant import shifting. This is not the same as state policies of import substitution, economic planning to locate new industries within a country. Quite the opposite in fact: import replacement is a self-generating economic process that is confined to growing cities. It involves firms in the city recognizing that they can make a product currently imported from another city. This is not simple replication of manufacture but a new production process utilizing specifically available skills and methods existing in the city. Hence, not a simple transplanting from one country to another as in state import substitution policy, but rather a customized adaptation into a new city. The outcome is the building of new agglomeration through drawing on a city's existing connections. And the latter leads to further changes: connections become adapted to the new situation (i.e. a loss of specific imports) by attracting new imports, thus generating import shifting. This happens because with the city no longer importing the replaced product, new import capacity opens up for different products. Jacobs claims this is not about one city winning at the expense of another because overall economic transactions have increased.

Her initial example, the development of a bicycle manufacture in Tokyo at the end of the nineteenth century, illustrates the key mutuality in economic spurts. Bicycles having been imported to Tokyo from European and America producers, local tradesmen developed bicycle repair businesses, often actually making replacement parts where necessary. As the scale of the latter business increased it reached an ability to construct whole new bicycles, not mass-produced like those imported, but adapted to the city's myriad workshops. Subsequently, a shift in imports towards new machinery, made in erstwhile bicycle exporting cities, further developed the process. The ultimate outcome in the twentieth century was that Japanese cities became the first engineering agglomerations to rival European-American industrialization in markets across the world.

This mechanism behind explosive city growth is the generic process that has created larger and larger cities culminating in today's global population having an urban majority. Historically this evolved through sequences of 'mother cities' exporting to other cities that then import 'replace and shift', thereby creating new agglomeration and connectivity. Two definitive examples of such economic development with world-changing effects are:

- Constantinople (as mother city) → Venice → Milan (and other Northern Italian cities) → Amsterdam (and other cities north of the Alps) resulting in Western European cities dominant in the early modern world economy.
- London (as mother city) → New York (and other northern US coastal cities) → Chicago (and other Midwest manufacturing cities) → Los Angeles (and other west coast cities) resulting in US cities dominant in the twentieth-century world economy.

In each case the second cities (Venice and New York) experienced explosive economic growth through replacing imports from the mother city, and this pattern continued down the chain to create large city networks that preceded today's world city network.

But there is nothing automatic about explosive city growth: cities decline when they are disconnected. Jacobs uses the case of classical Addis Ababa to illustrate this. An outpost of early Christianity, the rise of Islam in the eighth century through its conquest of Egypt left the city cut adrift from its city network in the Mediterranean. The result was many centuries of quietude. In recent times, disconnection has been the result of policies by states with anti-city agendas or else simplistic economic planning with little or no understanding of the relevance of cities. This is covered in Chapter 7.

Concluding supplement: agglomeration and connectivity externalities working together

It is important to understand that agglomeration and connectivity advantages operate best when in tandem. The geographical shift in the woollen textile industry at the beginning of the Industrial Revolution in England is a good illustration of this.

In the early modern period until the beginning of the eighteenth century, woollen textile manufacture was concentrated in relatively small cities in East Anglia and southwest England. By the end of the century it was centred in Yorkshire: Leeds, Bradford, Halifax and Huddersfield were new booming cities ushering in the Industrial Revolution in northern England. How did this relocation of a whole industry occur? Although we think of the Industrial Revolution developing through the introduction of the factory system, this was not a factor in this case because the rise of Yorkshire textiles preceded the adoption of large factories. Instead, we need to look at differences in how the industry operated between the old and new centres of the industry (Wilson 1971; Gregory 1982).

Starting with the industry in the more southern regions of England: these were traditional woollen industries with long-standing agglomerations of skills and practices that generated a product for the market. This product was then taken over by general merchants from London who marketed it in Europe as part of their established trading practices. This arrangement operated over many years and enabled a steady reproduction of the industrial agglomerations but without any notable economic expansion.

In Yorkshire, the cities also had their industrial agglomerations producing woollen textiles but the marketing was completely different. There was no reliance on outside merchants; local merchants were used. Even more important, they were not general merchants, they specialized in the marketing of textiles. Hence whereas London merchants were selling textiles as just another commodity to buy cheap and sell dear, Yorkshire merchants knew their cloth and how to expand sales in Europe. They had a deep knowledge of the commodity they were selling, which enabled them to be 'active merchants' searching out new markets and relaying market changes back to producers.

The result was that textile production in the two more southern English regions stagnated as agglomeration externalities went unsupported by connectivity externalities, whereas in Yorkshire the cities boomed through agglomeration and connectivity externalities working together.

City Insights E: T. H. Lloyd's German Hanse in England

According to T. H. Lloyd (1991) in his *England and the German Hanse 1157–1611* it was Boston, a relatively small city in Lincolnshire, that was 'the centre of the trade' for the Hanse in England in the early fourteenth century (p. 39). Strangely, and only temporarily, it was more important than London in the Hanse network of cities that straddled all of Northern Europe. The Hanse (commonly known as the Hanseatic League) defined itself as 'a firm confederation of many cities, towns and communities for the purpose of ensuring that business enterprises by land and sea should have the desired and favourable outcome and that there should be effective protection against piracies and highwaymen' (p. 7). Notice that the functions listed – enabling and protecting commercial activities – are what we expect modern states to carry out. The German Hanse provides a historical alternative to territorial governance, a city network. Lloyd's study is a unique contribution to understanding cities because it deals in detail with relations between an active network of cities, and a territorial state, the medieval and early modern Kingdom of England. It provides sight of a different form of city/state relation.

The Hanse lasted for nearly half a millennia, encompassing a long slow decline; in this discussion I focus on its early vibrant years (thirteenth and fourteenth centuries). The Hanse was an alliance of hundreds of German cities combining three main groupings: the Lower Rhine cities notably Cologne and Dortmund, the central cities notably Hamburg and Lübeck, and the Baltic cities where Danzig dominated. Major decisions were made at Diets held in Lübeck. Beyond German lands there were communities of Hanse traders in other cities across Northern Europe. These German communities were organized as kontors, which were recognized as integral parts of their host cities. In London the kontor was known as the Steelyard and illustrates well the incorporation of these communities of German traders into city life: the Steelyard had its own London Alderman and was responsible for financing and controlling one of London's gates, a security task no less (pp. 21–2). England's North Sea coast was part of the Hanse's trading realm and there were smaller kontors organized in Hull, Boston and Yarmouth, as well German merchants operating out of Newcastle, Lynn (now Kings Lynn) and Ipswich without kontor status.

The leading English kontor was in London because it was the seat of government. Therefore the terms by which German merchants operated in England were negotiated between the Steelyard, the king and his council and the Lübeck Diet. Other interested parties were the English merchants represented in the Parliament, and more particularly in the London Mayor's office. There is no sense of a 'national interest' in the negotiations, so that the various English kings did not simply side with the English merchants. For instance, German merchants often bypass the Mayor and appeal directly to the king (pp. 23-4); on one occasion this led to the king suspending London's city charter during negotiations (p. 21-2). Also, Parliament supported the Hanse in one dispute with English merchants (p. 45). Clearly this is all very complex, with commercial and financial interests interlocking in different ways. There are two overriding concerns. From the perspective of the king, the trade the Hanse brings to his realm can be taxed to produce an important source of income separate from his traditional dependence on the wealth of the country's main landowners. Taxing the Hanse provides a degree of fiscal independence from Parliament. The latter will also contain members whose fortunes depend on the Hanse. However, there is still an understanding that because the Hanse are foreign, a second common concern is for English merchants to have the same trading rights in German ports that the Hanse have in England. This reciprocity demand is English merchants attempting to join the Northern European inter-city network created by the Hanse.

Being a network process the English cities should be viewed through their inter-city links and it turns out that these were relatively simple: each east-coast city had a major trading partner. These city-dyads were: Boston–Bergen (Norway); Hull–Stralsund (Baltic); Ipswich–Cologne; Lynn–Bremen; and Yarmouth–Hamburg (pp. 85–91, 368). Cologne was London's main Hanse partner. The London Steelyard represented the collective interest of Hanse merchants in England but the east-coast communities had 'a considerable degree of autonomy' (p. 37). The provincial centres 'had little or no interest in the economic life of the capital' (p. 37); their commercial interest was a network focus, towards their dyad partner. It was in these circumstances that in the second half of the thirteenth century the growth of Hanse trade in England is greater in Boston than in London.

But this was not to last; by the end of the fourteenth century 'the London Kontor overtook that of Boston in importance' (p. 75). Two processes are operating here. In terms of connectivity, London always had more links than just to Cologne, and furthermore its trade encompassed a greater range of imports and exports than the east-coast ports. This meant that London merchants 'had greater capital and specialised knowledge of markets' (p. 79); in contrast, in Boston it was Lübeck merchants who actually carried trade to Bergen (p. 84). In terms of agglomeration, London was always the largest city, and furthermore when English trade moved from exporting raw material (wool) to manufactured goods (cloth) the latter production, and therefore potential import replacement and shifting (p. 97), did not occur in the port cities themselves. Thus 'there is no evidence of any deliberate attempt to discourage trade in the provinces. The decline was the result partly of commercial forces and partly of accidents' (p. 368). The latter refers to examples such as Hamburg merchants leaving Yarmouth in the wake of a corruption scandal (p. 163) and Boston hurt by England-Lübeck quarrels (p. 368). These examples simply illustrate the failure to build up economic resilience – building commercial links alone is not enough.

Finally, one ongoing bone of contention in negotiations between England and the Hanse is that the former were never able to find out the actual membership of the latter (pp. 36, 180). Such slippery flexibility can never be available to territorial diplomats; land is so very tangible compared to networks.

5. Demanding cities

Introduction: cities shaping landscapes

In this chapter and the next I am going to argue that cities generate economic inequalities. This proposition is not commonly recognized by those who study economic inequalities. In statistical exercises comparing numbers of poor people to numbers of rich people and all those in between, cities are treated as one specific scale of analysis; measures of inequality are more commonly carried out within countries and between countries. In this context the degree of inequality to be found in cities is interpreted as reflecting the relevant national and international levels; cities are viewed as encompassed within processes operating at these larger scales. The latter are by no means unimportant but such thinking completely misses the way that economic inequalities are produced. It is the basic dynamism of cities that is at the heart of inequality, urban, national and international. Below I consider creating inequalities as another component of the external relations of cities before focusing on inequalities within cities in the next chapter.

Previous discussion of cities as an economic process has largely dealt with production of goods and services, what cities supply for the market. But every supply requires a demand, and cities are also, very obviously, centres of consumption. In fact cities are incredibly demanding. Returning to the first large city network in Mesopotamia 5,000 years ago with its estimated population of over a quarter of a million urban residents, this constituted a massive market just for its routine reproduction. In economic terms, cities generate a completely new level of demand. This process is repeated with the creation of all civilizations resulting, for instance, in the irrigation of valleys, the terracing of hills and the trading of all necessary commodities that cannot be sourced locally. The key point is that this is the way cities mould their surrounding landscapes to varying degrees depending on the size of the demand.

In addition, the economic spurts that are the special feature of cities ensure that this landscape making is itself dynamic, not just reflecting the vibrant city process but being integral to it. Cities are continually changing the world around them. However, unlike the inter-city network process of the last chapter, these city external relations are strictly hierarchical: city demand calls the economic shots. Producers outside the city have to continually negotiate this city demand as markets respond to the changing consumption patterns of the city. This is the subject matter of this chapter, which is treated in two ways: first, the practical and logistical responses to the growths of city demand, followed by its extreme expression as landscapes of monoculture.

Beyond markets and fairs: centres of consumption

Cities generating economic demand did not happen as a fully functioning process: I have previously noted that the key mechanism of trade long precedes the existence of cities. As trade becomes more organized it becomes centred on safe places to buy and sell that may develop into city networks. However, with initial demand being relatively low there is insufficient trade to warrant permanent market centres. Therefore the selling and buying is time limited: local trade takes place in periodic markets, usually one day a week, and non-local trade is conducted in large fairs commonly annual in frequency. The former are commonplace across the world and continue in poorer regions. The latter, dealing with long-distance trade, is much more city-like and continues in specialist fields in modern cities: the Frankfurt Book Fair and Milan Fashion Week are well known, successful examples.

It is important not to interpret early organizations of trading as part of an evolutionary model of the development of cities. Great fairs have become associated with particular cities but there is no automatic progress from fairs to cities. The Champagne Fairs of the twelfth and thirteenth centuries in Western Europe illustrate this point very clearly. Champagne county was midway between the emerging commerce of Northern Europe and the more developed commerce of Mediterranean Europe: for a short time it became the key hub in the economic development of Europe This is where long-distance trade connections were made between merchants from both regions exchanging a wide range of goods including textiles and spices. There was an annual programme of fairs, six in all, held in four relatively small towns in the following sequence: Lagny-sur-Marne, Bar-sur-Aube, Provins, Troyes, Provins, Troyes. None of these towns became vibrant cities; being a centre of exchange of non-local products does not of itself provide a recipe for city development. Without local agglomerations of production and consumption these little towns were to be simply bypassed as European cities subsequently prospered through more direct connections, especially by sea between Genoa and Bruges. It is in Northern Italy and the Low Countries where the city process was stimulated. The Champagne towns, as mere locales for satisfying demand for goods elsewhere by production from elsewhere, were important trading hubs for a short period but never became network nodes, aka demanding cities.

In total contrast, the great cities of history have been immensely demanding, exceptionally so when their populations topped a million. This was the case with two great imperial capitals, Rome and Beijing, which will be used to illustrate the resultant shaping of faraway landscapes. In the former case part of the solution for feeding the people involved free distribution of grain to citizens. The logistics involved construction of a special port to cope with a continuous flow of ships from North Africa, especially Egypt. After incorporation into the Empire, Egypt became the breadbasket of Rome, economically structured for producing grain to satisfy one city's demand for food at the other end of the Mediterranean Sea. Satisfying Beijing's demand for food was even more spectacular: it required the construction of the Grand Canal, a system of waterways, both rivers and canals, traversing over a thousand miles linking the Yangtze and Yellow rivers. This enabled transportation of grain from the fertile Yangtze delta region to feed the people of Beijing. Although these two enormous logistic efforts are the result of imperial policies supporting their respective capital cities, the actual activation derives from the city process – agglomeration and connection – creation of a huge economic demand whose fulfilment was simply a necessity.

Imperial Rome and Beijing are extreme examples of city food demand requiring non-local solutions. However, all cities soon outgrow supplies from their immediate hinterlands and this is especially true of modern cities where local food is relegated to a niche market in an economics of worldwide market supply to a collective demand of ever-growing numbers of very large cities.

Supply regions: the dependency of specialization

Historically, cities with populations over one million have been exceedingly rare. Before 1800 they can be counted on the fingers of two hands: as well as imperial Rome and Beijing there have been other Chinese capitals – Kaifeng, Hangzhou and Nanjing – plus classical Alexandria and the most successful Islamic caliphate centre, Baghdad. Today there are estimated to be about 600 cities with populations of over one million. And of course, there are concomitant increasing numbers of cities exceeding half a million, and of over a quarter of a million, and so on. Hence the oft-quoted fact that, since the beginning of the twenty-first century, city dwellers constitute a majority of humanity. Satisfying this total city demand has immense environmental implications that we focus on in the final chapter; here, I show how we got to this situation.

In the first decades of the nineteenth century London reached a population of a million and overtook Beijing as the largest city in the world. This symbolic changeover was consolidated by other Western European cities, plus northern American cities, growing rapidly over the following century so that by 1900 contemporaries recognized they were living in a unique era of great cities (Weber 1899). Although more commonly referred to as the 'Industrial Revolution', thus emphasizing the production of new urban agglomerations, it was the consequent economic demand that transformed the rest of the world. In effect, the productive yields of numerous regions across the world were reoriented towards the booming cities of the North Atlantic zone. In this process the complex city economies of the industrial core zone generated specialist regional economies across the rest of the world, typically focusing on a single commodity to satisfy the needs of the new industrial cities.

The resulting commercial world geography was a global mosaic of places each associated with supply of a specific commodity. The list is lengthy but here are some major examples: Argentina/beef; Australia/wool; Bengal/hemp; Brazil/coffee; Ceylon (Sri Lanka)/tea; Egypt/cotton;

Malaya/rubber; New Zealand/lamb; South Africa/gold and diamonds; West Africa/groundnuts. At the same time the industrializing cities of northeast USA were generating their own domestic supply regions within the country, again specialist regions known for their products: Corn Belt, Cotton Belt; Texas/beef; Virginia/tobacco. The outcome was a highly integrated economic world divided into two unequal parts, one of complex city economies doing the demanding and the other of simple regional economies doing the supplying. This operated through most of the twentieth century and was characterized as a series of paired counterparts: initially termed developed/underdeveloped, later revised more optimistically to developed/developing, and then more neutrally to North/South. Running parallel to these terminological developments there was an alternative core–periphery way of thinking with a more relational approach. Instead of focus on simple classification by economic differences, emphasis was diverted to the interactions between them. The specific relation emphasized was dependency: production in specialized supply regions tracked the vicissitudes of demand in the cities of the economic core zone. Thus, through emphasizing the dependency process of making inherent inequalities, the core/periphery approach treated the commercial relations between places rather than just focusing on the contrasts in their economic circumstances.

Dependency is not a permanent relation. It is produced in the first place as commercial opportunity to satisfy consumer demand in faraway cities. The process requires development of an operative logistics that require initial hubs for transfer of commodities. But the latter can expand to become emerging cities in their own right, albeit economically dependent. Hence every supply region has its main port or land gateway. It is the efficacy of these incipient city processes that determines the degree to which economic dependency is enduring. This is a matter of converting the dependency relation into inter-city mutuality as described in the last chapter. And since logistic hubs are specialist clusters of activities, it follows that developing inter-city mutuality is contingent on the development of a nodal economic agglomeration.

In highly segregated colonial cities the mixture of local indigenous networks and long-distance political networks is a recipe for maintaining dependence. At the other end of the spectrum there are logistic hubs like San Francisco for the mid-nineteenth-century Californian gold rush, and Chicago as the railway hub for moving Midwest meat to the east-coast cities and beyond. In both US cities import replacement and shifting took place on a large scale to generate increasingly more complex agglomerations, thereby enabling these to become booming cities and take their place in an emerging coast-to-coast American city network. But overall across most of the world the mechanisms for maintaining economic dependency were stronger than the development of inter-city mutualities through into the second half of the twentieth century. This is the worldwide inequality legacy upon which a new powerful economic globalization developed in the late twentieth century to produce our more complicated global economy today.

The landscape expression of dependent specialization is monoculture. This arises from the market rewarding efficiency in production, which is best achieved through producing a single commodity. Originally typified by plantation agriculture – for sugar, then cotton and tobacco – it subsequently became typical of all supply regions. Today such production has become known as 'industrial farming' with monotonous landscapes of single crops and grazing. The use of the adjective 'industrial' is perhaps ironic in this case because this is an economic process that is not a respecter of production sectors. Large-scale specialization of manufacturing also produces economic monocultures in cities: Manchester as 'Cottonopolis' in the nineteenth century and Detroit as 'Motown' in the twentieth century are obvious examples. The problem is that such specialization in cities leads to the question of what to fall back on when the prime production becomes 'old work'. Answering this became specifically urgent in the second half of the twentieth century because of worldwide industrial relocation. As the traditional industrial centres of Western Europe and northeast USA rapidly declined they were replaced by production in previously non-industrial regions: initially southern US cities, East Asian cities and Latin American cities, and latterly Chinese cities that today constitute the world's main industrial goods supply regions.

These new economic circumstances make the previous terminology contrasting dual economic development positions much less relevant for describing cities in contemporary globalization. These dualities were conceived when the adjectives 'modern' and 'industrial' had effectively become synonyms – to be modern was to be industrial – an equivalence that lasted for more than a century. But with this no longer being the case the 'modern city' has had to be reinvented and in the process it is consumption rather than production that has come to the fore. Cities are even more demanding in the twenty-first century, being satisfied by the greatest 'industrial revolution' of all, in Chinese cities. A demand too far – this global story is the subject matter of Chapter 9.

Concluding supplement: whither the rural?

The power of economic demand through cities has been the main theme of this chapter. The nature of this form of power is subtler than how we normally understand the deployment of power. Reprising Chapter 1, power through taking action is what states do – passing laws, making policies, waging war are each expressions of this power – whereas the power dispensed through economic demand is more basic: it derives from the need to reproduce the everyday behaviour that sustains cities as economic process. I have used the need to feed city dwellers as having reshaped rural landscapes as supply regions but there is an even starker example of this power to change landscapes: the need to supply water for rapidly growing cities. Here the result is typically not a reshaping of rural landscape but rather its actual disappearance!

For instance, in the recent past both Manchester and Los Angeles have experienced looming water shortages and have taken similar steps to forestall the potential crisis. In England, Manchester's solution was to flood a valley 160 miles away in the Lake District; for Los Angeles, a valley was flooded in the Eastern Sierra Nevada 233 miles away. Both required political measures to destroy the valleys and build aqueducts to divert water to the cities. But these politics were essentially just practical mechanisms. The key point is that, despite local opposition in both valleys, Manchester and Los Angeles were going to get their water, if not from these valleys they would have had to flood alternative valleys. It is simply inconceivable for the two cities to run out of water causing the relocation of whole cities. Meeting their needs is a requirement, a necessary city demand satisfied.

This brings to the fore the meaning of the term rural. Usually understood as the obverse of urban, here I have interpreted it as a product of the city. As such it is best viewed as a type of landscape within a singular economic process that is city-making. Thus it is not a process in its own right. This has been widely recognized when societies in richer countries have been referred to as 'urban societies' even though the vast majority of land is overtly rural. Today this way of thinking has been further extended to the global scale: pollution from a hugely enhanced city process, from gas emissions to plastic waste, is found in every part of the Earth far beyond the archetypal rural to the uninhabited polar regions and the deep oceans. City process is everywhere.

City Insights F: William Cronon's Booming Chicago

Cronon's (1991) *Nature's Metropolis* is about one city's remarkable growth from a small settlement nestled around a military fort in the 1830s to become one of the great cities of the world by the end of the century. A well-trodden path, Cronon tells this story as several 'historical journeys between city and country' (p. 8): his purpose is to depict a dual transformation by giving equal prominence to the city's hinterland, the 'Great West' stretching across the continent to the Rockies and the Pacific. In his 'unified narrative' (p. xvi) the making of Chicago cannot be separated from the making of the Great West.

In this undertaking he directly challenges Frederick Jackson Turner's traditional 'frontier thesis' of American westward expansion, which we will all know through films, literally called 'Westerns'. Cronon replaces this inherently rural story with a metropolitan-led development process. Curiously, in making this argument he brings to the fore the land speculators with their booster language that is 'not read anymore' (p. 46); he suggests they have a more realistic understanding of the development of the west than rural-biased historians. For the speculators, the city 'radiates an energy' (p. 13) within a 'gravity theory' of cities (p. 38). This is Cronon's Chicago – a 'vortex' (p. 29) becoming an 'economic earthquake' (p. 247) with 'explosive growth' (p. 265). But for Cronon it was a double transformation, new rural worlds were simultaneously created. He describes Chicago as 'the site of a country fair, albeit the grandest, most spectacular country fair the world has ever seen' (p. 98). This explains what Cronon calls his 'peculiar' approach: to organize his book around the topic of commodity flows (p. xvi).

There are three main commodity flows articulated through Chicago that completely changed both city and hinterland: movement of grain, timber and meat enabled by the coming of the railways. All three are marked by immense quantities enabled by logistic innovations. In the case of grain, initially moved in individual sacks, the key innovation was the giant grain elevators. These required the control of inputs from multiple sources resulting in the strict grading of grains, thereby totally revolutionizing the marketing of the commodity. This in turn enabled financial innovation: conversion of receipts for grain to become capital flows (p. 120), which when linked to the new telegraph service (p. 121) developed into Chicago's futures market (p. 124).

The link between rapid agglomeration and colossal innovation is best represented by the meat industry, notably cattle. Unlike hogs with a tradition of cured products, cattle were slaughtered, butchered and consumed fresh so that live animals had to be transported direct to markets. All this changed with the invention of the refrigerator railcar and the disassembly line. The latter was a factory where each carcass was divided into multiple parts with little or none of the huge waste typical of abattoirs. It involved specific cuts of meat for the new meat-packing industry to sell frozen pieces to the ordinary consumer. Persuaded by intensive marketing, the American meat market was entirely revolutionized. But the carcasses supplied much more than meat: a single creature was turned into 'dozens then hundreds of commodities' (p. 250). And so meat-packing companies expanded into corporations with, for instance, their chemical laboratories producing buttons, brushes, glue, fertilizers, and with dried blood and crushed bone-meal becoming powders for druggists to dispense (p. 250). This is innovation in overdrive.

Chicago's hinterland was transformed – goodbye tens of millions of bison (pp. 237–41) – and the city became 'metropolitan' – new factories producing all manner of goods (pp. 311–12), accompanied by department stores where the growing population could buy them, plus high culture – theatres, orchestras, art galleries – and a large publishing industry (p. 281). This last led to another transformative innovation: mail order catalogues from which people in the hinterland could consume metropolitan goods: Montgomery Ward and Sears created this new market (pp. 335–7). Chicago had indeed become a 'busy hive' (p. 337).

This story of immense economic success was premised on Chicago's position in the new railway network that was integrating the American economy. The city's connections formed a trunk and fan pattern (p. 90). Tracks fanned out from the city to the west, the hinterland, but to the east there were just mainline trunk routes. Both railway nets terminated in Chicago (p. 83). This made it a 'gateway city', neither west nor east but in-between (p. 283). Crucially, relations between these two directions were very different, to the west supply regions, to the east market demand that made possible the whole city growth process. Thus, although Cronon claims to be integrating city with hinterland he is actually doing much more.

Throughout the book he is linking Chicago into the broader world economy. For instance there are references to: the 'metropolitan core ... presumably somewhere off to the east – whether on the American or European side of the Atlantic' (p. 51); Chicago is 'the western outpost of a metropolitan economy centred on the great cities of Europe and the American North East' (p. 60); 'dressed beef brought the entire nation – and Great Britain as well – into Chicago's hinterland' (p. 238); and 'the world must become Chicago's hinterland' (p. 255). The term hinterland in the last two quotes is inappropriate because, being part of this wider network of cities, relations to the east have a more equal mutuality than relations to the west's supply regions. Cronon argues that without New York, with 'the most powerful financial institutions of any city in North America' and with 'the most direct access to European markets', there would have been no Chicago success story (p. 62).

One final point, Cronon's previous books had been histories of rural places, and coming from this position he has produced what I consider to be the best book written about a city. He understands cities because he knows the rural.

6. Divided cities

Introduction: migration fuelling inequalities

Cities need migration. Until the public health initiatives of the nineteenth century the dense living of cities was especially unhealthy. The result was that death rates exceeded birth rates. Hence without the input of migration past cities would have simply withered away. But they didn't: the attraction of cities generally outpaced the high mortalities. Additionally migration was necessary for more than simply cancelling out the toll of diseases, economic development through cities required evermore labour. Hence a key addition to the demands of cities described in the last chapter is a growing demand for new labour. Thus, dynamic cities depend on migration to meet an incessant creation of new work. It is the latter process that has continued to operate after the public health reforms of the nineteenth century reduced death rates. As city economies have collectively grown at ever faster rates so too has the demand for labour. More people are migrating to cities today than in any other historical era.

Migration does more than maintain or grow the stock of labour in cities; it fuels inequalities. The main sources of migrants into cities are from labour supply regions, traditionally identified as rural peasants who have agricultural work skills. The move to the city essentially deskills them. Therefore they search out the work that requires little or no prior training and therefore pays irregular and low wages. Thus migration continuously provides the labour at the bottom of the employment scale, such as informal work hawking cheap goods or more formal work in unregulated factories. This is the mechanism that is fuelling economic inequalities even as cities generate economic development.

There is a scalar paradox in this process: while fuelling inequalities within cities, migration reduces inequality across the world. How can this be? Remember that for migrants, cities represent economic opportunities. The work they do in cities should therefore be better – more remuneration, greater prospects – than the work they left behind. And the city is also getting a good deal. Although many migrants send money home to their family back in the supply region, a loss to the city economy, this is more than compensated by the supply region bearing the costs of producing the labour in the first place. Subsistence, health, education and all that produces people for the labour market is taken care of far away from the city: it gets its new labour for free. The crucial point is that the result of there being a higher proportion of people living in cities is that overall inequality is reduced, even though it is being maintained within cities themselves. This paradox is expressed today by economic inequality reducing globally, largely due to the massive rural–urban migration in China over the last four decades, while immense poverty endures in cities across the world.

Spatial structures of cities

Economic inequalities are manifest in the spatial structures of cities. Traditionally, the most desirable places to live in large cities were in the central areas, with undesirable land uses and their associated workers banished to the outskirts. For instance, tanning of hides in leather trades produced a noxious environment and was to be found on the edge of the city, often outside the walls. This common spatial pattern was reversed with the rise of the industrial city in the nineteenth century where the huge economic growth and its associated factories and workshops created a new city focus. This famously smoky world of industries surrounded by dense housing led to the movement of the more affluent away from the dirt and grime in early examples of public transit suburbanization. Thus were most city dwellers left behind in immensely overcrowded places, at a new scale of squalor that required the invention of a new word: slums.

Hence, as well as exceptional economic development, the nineteenth century also bequeathed novel social problems to the twentieth-century city. The need to eradicate slums led to multiple political and social reforms, with associated researches and new practices, most obviously city planning. New understandings of cities were required and these were supplied by sociology researches at the University of Chicago in the 1920s and 1930s. Their most famous product was Ernest Burgess' (1925) concentric model of city structure based upon their home city. The model consisted of a central business district (still known as the CBD) at the centre of four surrounding zones in the following order: (1) a blighted transition zone mixing factories with poor housing; (2) a ring of improved housing for workers in more stable occupations; (3) a middle-class housing zone; and (4) the commuter zone of dormitory suburbs. These represented a very dynamic city whereby new migrants got their foothold in the blighted zone, the next generation attained housing in the more stable zone, and after which the two outer zones became attainable.

This iconic model of the city with economic status graduated upwards away from the city centre dominated much of twentieth-century thinking on the city. Both simple in form and dynamic in practice its relevance was advanced in two distinctive ways, one in theory the other in practice. The former was an economic theory of land uses where bidding for housing amenity against business accessibility results in median family incomes increasing with distance from the centre. But even more impressive was the actual post-Second World War boom in suburban housing development that confirmed a clear inner city/outer city dichotomy based upon economic inequalities but in a new world of rising incomes and consumptions. The latter ushered in a more affluent world so that suburbia was far more broadly based than in the original model, a spatial structure owing more to Los Angeles than Chicago.

This was a new, more expanded spatial structure enabled by the automobile and its highways. No longer suburban islands around public transport stops, suburbia-by-car filled in the gaps to produce a new urban landscape of affordable housing for workers. But it was more than a physical landscape, it required a way of life focused upon consumption, not just the necessary car but a wide range of products to facilitate home living: fridges, washing machines, vacuum cleaners, lawnmowers, televisions, etc. And the production of these commodities – new work – provided the economic basis of the post-war boom in America's cities. Furthermore, US economic success attracted foreign imitation generating an 'Americanization' of others' ways of living: suburbanization and its consequent consumptions became the growth pattern of cities across the world in the second half of the twentieth century.

This spreading out of cities around a continuing inner city of older housing and CBD took different forms in different countries. In the USA many suburbs were politically separate from the city government, thereby avoiding contributing taxes to the city's budget whilst still enjoying the cultural offerings of a big city – theatres, opera houses, concerts halls, museums, etc. In many European countries there was much more suburban public housing, thereby partially ameliorating the economic inequalities between inner and outer city. But this spatial differentiation was never that simple: as the spreading out continued it often resulted in joining together neighbouring cities to produce large city-regions with multiple cores, which are discussed in Chapter 8.

Partly because suburbanization led to the loss of people and with them finance (i.e. a tax base) from big cities, it contributed to the widespread crises of cities common into the 1970s. The revival of cities came with the rise of economic globalization wherein cities became the key centres of corporate organization. This is reflected in a remarkable growth in high-rise office blocks – skyscrapers – that mark city centres in cities across all parts of the world, and which, literally, dwarf the old CBD structures. This change has created a new pattern of inequalities. The corporate economy has generated a band of highly remunerated personnel with huge consumer demands that have, in turn, generated low-income jobs – domestic, hospitality – to meet their needs. At the same time, the huge agglomerations of these cities have been instrumental in creating large numbers of new, medium-waged jobs. These changes have resulted in a reversal of the urban dispersion by economic status; a movement of the rich back towards the centre.

This new process, the obverse of suburbanization, is called gentrification. Large cities did maintain some high-income enclaves, for instance London's Belgravia and New York's Upper East Side in Manhattan, but gentrification produced a whole new broader pattern. We are back in the inner city, the first ring in the concentric city model, but which is now deemed to be attractive. A zone for 'slum clearance' into the 1960s, housing that survived this physical onslaught is attracting high-income households. Explanations for this change-around take two forms. The economic opportunity argument is that these places are recognized as new investment openings: property bought up cheaply to be sold or rented at a large profit. The consumption argument is that the young people with the new jobs in the reviving cities wanted inner-city living as part of a cultural consumption shift away from suburban living. These two explanations are not incompatible; both have been shown to be operating. The result is the production of a new urban space involving multiple upgrading of houses plus conversion of old neighbourhood stores – grocers, bakers, butchers, etc. – to restaurants, bars and coffee shops. The key change is displacement of poorer original households by new richer households. Thus, gentrification is a process where the differential power created by economic inequalities operates to transform the spatial structure of the city. This power process has been accentuated by globalization: in some parts of major cities there are the very rich who are 'super-gentrifiers', buying up the most expensive housing as part of a global property market across the leading cities of the world. This is discussed further in Chapter 8.

Cultural kaleidoscopes

Economic inequalities are only half the story of divided cities. Material differences amongst citizens intersect with the cosmopolitan nature of cities: spaces of income and wealth distinctions are commonly cultural segregations of people. This is an integral part of the migration process with which this chapter began. Migrants do not normally just turn up in a city lost in its immensity. Their goal is to choose a city, and within the city a neighbourhood, where others they know have gone before them. These may be family, people from their home village or town, or more generally a recognized community of their ethnicity. Thus in a rapidly growing city with a range of labour supply regions the result will be a patterning of ethnic enclaves. For instance, in 1920s Chicago the first inner ring of transition included a neighbourhood called 'Little Italy', where the language and traditions of the homeland were continued: new migrants could feel at home.

In the Chicago growth model it was posited that these ethically distinct Americans would gradually assimilate into their new country's norms through the generations so that the outer rings would be far less segregated. For instance, third-generation Italian Americans prospering in the suburbs of 1950s Chicago would likely not have a working knowledge of the Italian language. This is how American cities were expected to create regular American citizens. And it worked to a degree. The great exception was an internal migration: the American South as a supply region of African-American labour for northern US cities. Instead of the resulting ethnic enclaves becoming conduits to the suburbs, they became seen as 'ghettoes', a form of bounded segregation consequent on so-called 'white flight' from the inner city. This is where American residential integration was stymied: in the 1960s the outcome here was riots not suburbs.

The limitations of cultural assimilation through cities were not just a feature of the USA. Degrees of cultural and physical differences combined with the nature of the dominant city population resulted in varying levels of segregation: the limiting case was in South Africa where the policy was actually to create ghettoes of black African labour in the Apartheid city: Soweto (named from <u>So</u>uth <u>We</u>st <u>T</u>ownship) on the edge of Johannesburg became the largest of such settlements. But such deprived and coerced populations still constituted agglomerations and as such generated new ways of working and living. Famously in the USA these included popular music genres emanating from New Orleans, Chicago, Detroit and Harlem, New York. With such cultural differences in cities creating their own identities the whole idea of assimilation came under scrutiny. Instead of eradicating difference, diversity, the city as cosmopolitan meeting place, can be something to celebrate.

Living within a diverse population means that neighbours can be viewed as strangers. As such they are less predictable than friends; they think and behave differently. There are two contrasting reactions to this situation: to shy away, keep your distance from the dangers they pose, or to embrace the opportunity of experiencing new ways for enriching lives. The former leads to strict segregation – streets, parks, shops all identified within marked territories – as reported above. With the latter, however, instead of the opposite of segregation being assimilation and resultant loss of traditions, there can be respect for different traditions through positive everyday encounters and building joint community initiatives. Hence the divided city can be marked by cosmopolitan vitality.

To reach such a positive outcome requires people having a continuing confidence in their own traditions. This is now made possible by contemporary communication technology. Whereas migrants in Chicago's Little Italy had largely burnt their bridges with their homeland bar a few letters, today's media enables immediate contact with family left behind and homeland in general. Rather than ethnic enclaves there are cultural diasporas: transnational communities of common ethnic, religious or national composition. These cultural inter-city networks are today's context in which new migrants relate to people of different backgrounds across their chosen city. However, it is too soon to know whether this globalization of cultures and ethnicities will create a positive cosmopolitan effect on cities or if what were once local conflicts become adjuncts of international conflicts.

Concluding supplement: cities and revolution

Cities are unruly places. Rapid growth of large numbers of people from diverse supply regions creates a recipe for social unrest. This can take multiple forms but at their most extreme they become revolutions in cities and revolts and rebellions in supply regions.

Paris is the modern city of revolutions, with major political upheavals in 1789, 1830, 1848, 1871 and 1968. As a capital city Paris encompasses a political agglomeration – state institutions representing domestic politics plus diplomatic zones representing wider political connections. This concentration of political practice, thinking and enterprise has an innate potential for radical innovation, which is what each revolution offered with varying degrees of success. Many of the '-isms' of modern democratic politics – nationalism, liberalism, socialism, anarchism – have their major initial stimuli in these revolutions and continued to develop in subsequent revolutions. Of course Paris was never alone, with other cities involved in France and beyond. This was especially the case in 1848 where revolutions occurred in capital cities across Europe, sometimes termed the European 'Spring of Nations', a contagion replicated over 150 years later (in 2010) with the 'Arab Spring' of uprisings in North African and Middle Eastern capital cities. In short, capital cities provide not just the spatial settings for sporadic political revolutions but, more fundamentally, the political raw material.

Most cities are not capital cities, but this does not lessen their innate potential for unrest leading to fundamental change. Away from the formal political arena, economic conflicts abound both between capital and labour and within capital. Thus the mass labour demonstration in Manchester in 1819 that became the 'Peterloo' massacre and the

Haymarket riot in Chicago in 1886 (leading to May Labour Days) both became remembered as milestones in the advance of Labour as a political force. But busy industrial cities also advanced business interests with political innovations: hence in Victorian England it was 'Manchester Liberals' who propagated free trade economics countered by Birmingham's tariff reform movement. These were new politics created in specific city contexts and diffused to other cities in Britain and beyond.

Whether searching for fundamental change in political rule or changing economic relations through city agglomerations and connections, these developments are obviously matters of state and can only be understood through the prism of city/state relations.

City Insights G: Hsiao-Hung Pai's Stories of China's Rural Migrants

It is commonplace to note that the Chinese economic transformation of the last four decades has been made possible by the largest rural–urban migration movement in history. It is much less noticed that each of these millions of workers is an individual with hopes and expectations that do not usually match their experiences. Hsiao-Hung Pai is a UK-based journalist from Taiwan who has chronicled the lives of migrants in her book *Scattered Sand* (2012). This phrase is widely used to describe the mass movement as an unorganized shift of anonymous labour to the cities. She reveals the humanity that constitutes the scattered sand by telling the stories of migrants in their own words.

I have selected two of the 'grains' she so vividly reveals as ordinary people caught in extraordinary circumstances. I focus on their largely negative work experiences, the miniscule possibilities of redress from public authorities, and the subsequent state of their well-being – both physical and mental.

Peng is a young farmer from a village in Liaoning province, the only child of his widowed father, who migrates at the age of 17 because the family's very small landholding can no longer maintain even a household of two. He moves to the nearest city, Shenyang; Hsiao-Hung Pai meets him looking for work in the city's Lu Garden Labour Market where thousands of migrants congregate every day in search of jobs. Peng's intermittent work history is as follows: security guard in a small hotel, sends two-thirds of his income back home but is dismissed without notice after two weeks; temporary work as a labourer on a building site for two days; another security job at a local brewery at lower pay, fired when queried his pay, he comments 'Bosses can do anything they like' (p. 20); jobless for a month so returns home but sent back to Shenyang after two weeks; back at Lu Garden Labour Market but no luck – his friend commits suicide (p. 29); returns home but father forces him to leave after two weeks – he feels 'I am just like his working buffalo' (p. 30); recruited at Lu Garden for security job in Beijing, very excited but, unknown to him, the firm he is to work for, Tianhe Antai, is a criminal organization; in Beijing, ID paper is taken, he's sent to guard an insurance business but is not allowed to leave premises, escapes after three weeks, no payment but retrieves ID – there is a saying ,'You can get work from Tianhe Antai, but you can't get money' (p. 34); now 'utterly demoralized', because 'in practice ... the law has made no difference for workers like Peng' (p. 35), he returns once again to his village; leaving again, no luck at Lu Garden market so heads back to Beijing – 'I am determined to find something' (p 35); recruited for security job at a hotel, very low wages – 'You take it or leave it ... stay like a slave, or go back to the countryside – who gives a damn about you?' (p. 37); not a living wage so forced to move on, back to the local labour market and luck changes, a security job at the Golden Sail Holiday Hotel, celebration, he can now send money home to his father. His final words to Hsiao-Hung Pai were: 'I hope that when you return to Beijing one day, you will see that I have got myself a more senior position at work and have done something better with my life' (p. 40).

Liu Min is the son of rural migrants from Sichuan but still refers to himself as 'from the village' even though he has spent his whole life migrating for work from the age of 17 (p. 197). His work chronicle is as follows: a brick-burning job for one year in Shanxi, wages sent home but 'thrilled' to be independent (p. 198); recruited as a gold-digger in Gansu's mountains, harsh conditions with good money but quit after 40 days; goes to Guangzhou but no job – feels 'utterly defeated' after borrowing cash to get home (p. 201); goes to local city, Chengdu, rents a motorbike and works at the station as a taxi for two months; back to Guangzhou, this time 'ready to make a serious effort to find work' (p. 202) he started as a motorcycle taxi without knowing the city or speaking the local language; girlfriend finds him a job in a garment factory – 'a real job' to be celebrated (p. 203) but laid off in second month; gets a warehouse job stacking shelves – piecework with policy that 'all workers owed two months' wages and paid in third month' (p. 204) so had to keep taxi job to make ends meet; not wanting a life toiling in a factory he returns home to set up general provision store but the 'family business' fails; back to Guangzhou and another garment factory, no longer needed motorcycle taxi supplement, but 'he felt that his life had become purposeless'; laid off in the recession, owed two month's wages but not paid. Lin Min's dreams had been 'dashed by the ruthlessness of life' which he shared with a whole generation of migrants 'who had given the city their sweat and blood' (p. 206).

Peng and Liu Min are not alone; they are representative of the myriad work chronicles that Hsiao-Hung Pai provides. But they are not alone in another sense; large movements of Chinese are also commonplace as tourists, 'the footloose global trotter' representing the other side of

China's growing unequal society (p. xi). There could hardly be a greater contrast. And for the rural migrants it is institutionalized: they are officially *nongmin* which makes them second-class citizens not entitled to many state services (i.e. no pension) (p. 3) and treated with contempt by city dwellers as *mangliu*, meaning 'blind flow', implying directionless with no purpose (p. 5). No wonder Liu Min resents his origins: 'to be born into a peasant family ... was like having a brand on your body' (p. 201). But Peng is more assertive of his role: 'without migrant workers, Beijingers would starve' (p. 40). When he asks Hsiao-Hung Pai about London, he understands immediately when told that Canary Warf 'is guarded by hundreds of non-British security guards employed by dodgy agencies' (p. 40). Migrants earning peanuts like us!

7. Cities in states

Introduction: a modern paradox of cities

There is an odd inconsistency in the history of relations between cities and the external political entities that have authority over them. Although our world of myriad great cities from the nineteenth century onward is unique compared to all previous societies in its levels of urbanization, cities in these earlier worlds often had their own distinct political powers recognizing their economic role in society. For instance, in medieval Europe the phrase 'city air makes you free' had a literal legal meaning. A peasant migrating to the city was breaking his labour commitments to his lord, working the land to which he was legally bound as a serf. But escaping to the city and being undetected for a year and a day cancelled the duties back in the village; he was now a citizen, able to partake in the city economy. Thus were cities not just different types of settlement, they were special places that enabled the bypassing of traditional political hierarchies. But this legal exceptionalism was swept away with those traditional hierarchies in the development of our modern world. The curious result is that modern cities have none of the special status – formally respected for their different role in society – that their pre-industrial forebears possessed. This is a modern paradox of cities, being unexceptional locales within the sovereign territories of states despite their unique demographic and economic prowess. A recent dispute between a city and its state neatly illustrates the modern situation.

It is unusual for a piece of art to be so symbolic of a city's identity as that of Michelangelo's marble statue of *David* and the city of Florence. Commissioned by the Florentine government in 1501, as well as being one of the masterpieces of the Italian Renaissance it represents locally the resistance of little Florence against greater outside forces: the Goliath in this case usually being Rome. Today the statue is a massive tourist attraction, central to the city's economy. But who actually owns this wonderful piece of art?

The answer seems to be straightforward: Florence. But despite being commissioned by the city and being located in the city for over half a millennia, this understanding has recently been in dispute. In 2010 the Italian government in Rome claimed ownership of the statue. Their interpretation was also quite straightforward. In 1871 the establishment of a unified Italy meant that the newly formed state was the legal sovereign successor to the Republic of Florence. Thus the dispute centres on the legal concept of sovereignty, the conclusive political power over territory and all therein. Although the statue has always been in Florence, the key point is that Florence itself has changed: from being its own independent political entity to mere municipality within a larger national sovereignty. Put bluntly, in the modern world Florence has been politically downgraded.

Of course for most people this political bickering is inconsequential, *David* is to be enjoyed not disputed. But it does highlight the question of the relationship between cities and the states that contain them. Unlike Florence, most cities have never had independent status but like Florence their encompassing state does now define their physical identity. And this is a problem because cities and states exemplify two different spatial logics: as noted previously states are territorial and create a world of frontiers and boundaries, whereas cities function through networks and connections. Seeing cities through the lens of the state is to simplify their being: each is a patch of territory that requires administration. Thus are cities endemically misunderstood by their states in the normal conduct of state practices; the essence of cities in agglomeration and connectivity is typically ignored in the neat defining of manageable areas for local government. However, incongruously, this all changes in times of war. Existential threats to states bring cities to the fore as crucial assets at home, and assets of the enemy state to be targeted: both agglomeration and connectivity become central to state wartime policies, domestic and foreign. In this chapter we investigate the city/state relation in both contexts, routine lack of understanding and existential acute understanding.

The debasing of cities

The first task of the state with respect to cities is to place them within the administrative map of the country. This is to bound them; the drawing of 'city boundaries' defines who and what is in the 'city'. The result has been that many cities are typically found to be under-bounded; parts of the city economy are deemed not to be in the 'city'. For instance, the 'city' of Florence in the dispute above is a commune within the Italian province of Tuscany and only includes about a quarter of the population that constitutes today's Florentine urban economy, the real city of Florence (i.e. its city process).

From the point of view of the state the problem with cities is that they grow, often very rapidly. This problem is generic for the simple reason that, as previously argued, it is the very nature of cities to develop and thereby enlarge themselves. Hence initial drawing of boundaries to demarcate a city – typically around the built-up area – to define its administrative area is soon outdated as the city population spills over into adjacent areas. Over time, the state's formal designation of the 'city' on its administrative map can become an anachronistic relict of a different era. Coping with the problem takes several forms, none of which provide long-term solutions. Indeed they mostly exacerbate the situation, leaving cities circumscribed, their vital economic process operating independently from how the state organizes its territory. At their worst, state policies can become explicitly anti-city.

The reorganization of France's administrative structure after the revolution of 1789 is one limiting case of city/state territorial relations: cities were simply ignored. Cities were viewed as spatial aberrations to those drawing the new rational administrative map. With a 'one policy fits all' approach the diversity of cities with very different growth experiences was considered of no relevance for organizing political control of France's sovereign territory. The country was divided into 83 departments in a patchwork of compact, roughly equal-sized territories. This took no account of population patterns or ways of life: the dehumanizing of space was explicitly confirmed by departments being commonly named after physical features, such as rivers and mountains. Thus Paris found itself in the Department of Seine. Each department was governed from the town that happened to be nearest the centre, this was required for security reasons: a political centre that could be reached from the edge of a department on horseback in a day. Thus local government was not located in the biggest city where it was not central. Departments remain today, with some reforms and additions, to continue to provide France with a very geometric local administration map.

Given the role of cities in political revolutions as described in the previous chapter, it is hugely ironic that since 1789 other states created by revolutionary means have similarly downgraded cities. In communist states in the second half of the twentieth century the one size fits all approach is apparent in state economic planning. This total lack of understanding of how cities work was clearly represented by the USSR and the People's Republic of China. Both embarked on explicit anti-city policies, curtailing urban growth and thereby stymying consequential economic development. For instance, in the 1960s China's 'cultural revolution' led to many citizens being forced into farm work resulting in the country actually de-urbanizing. But the extreme case came with the successful revolution in Kampuchea in 1975, where the new state policy involved evacuating the country's largest city; citizens were literally marched out of Phnom Penh and forced to become farm workers. The result was more than a million people dying in the 'killing fields' of the country before the regime was overthrown in 1979.

The above examples illustrate an instrumental functionalism from a security perspective Although without the tragic outcomes, another functionalism based upon simple administrative criteria shows how states consider cities as merely vehicles for implementing national policies. This is clearly illustrated by the English local government reform of 1973, which is still largely in operation. In this case reform was seen as a problem of size of local government units, framed as a contest between democracy representing a need for smaller units and efficiency supposedly linked to larger units (Dearlove 1979) The latter consideration largely prevailed, opening the way to using management and organization theory to redraw the administrative map of England. This meant that many smaller cities were simply removed from the map for being too little while simultaneously some major cities were deemed to be too big resulting, for instance, in Liverpool and Manchester being severely under-bounded. In some cases medium-sized cities had to be combined to make up the right size, resulting in new local administrative areas having new names invented for them. For instance, two historic industrial places Halifax and Huddersfield became 'Kirklees', neutrally named after a settlement that no longer exists! Both Halifax and Huddersfield are very much smaller than Liverpool and Manchester, but combined they are roughly the same size as the two bigger cities in their reduced form. Thus Liverpool, Manchester and Kirklees appear on England's administrative map designated equally as 'metropolitan boroughs': a plain and simple state solution to annoying differences that always occur between real cities. This is instrumental functionalism taking no account of the actual places themselves, the state is seeing cities, large and small, as mere administrative cogs as if they had no purpose other than doing the central state's bidding.

But whatever the administrative framing of cities, on the whole they still keep growing. The simplest and most obvious method of accommodating urban growth has been incremental annexation of adjacent communities as they become part of the expanding city: what were villages became new urban neighbourhoods. This was commonplace in relation to nineteenth-century industrialization in both North America and Western Europe as many cities grew in unprecedented numbers at exceptional rates of growth. But this process came to an abrupt halt in the twentieth century in the wake of new suburbanization. As noted in the previous chapter, the new suburbs with their relatively affluent residents resisted incorporation in order to maintain control of their local tax returns. The result was most acute in the USA where 'doughnut' administration maps resulted with multiple, largely white, suburban municipal units surrounding a poor inner city, with large non-white populations. This was extreme political fragmentation of cities as economic process.

The reaction against this political urban fragmentation has been to call for a regional designation of cities. This had led to the promotion of city-centred regional planning in the first half of the twentieth century. Designated 'conurbations' by the visionary thinker Patrick Geddes, they combined multiple political units into economically coherent areas. He identified such city-regions around major cities such as Chicago, Paris and Berlin but also recognized multiple city zones such as 'Clyde-Forth' (Glasgow/Edinburgh) in Scotland and the Ruhr (Düsseldorf/Dortmund/Essen/Cologne) in Germany. These ideas were particularly influential in New York and London. In the former, the Regional Plan Association of New York, founded in 1922, set about proposing policies for a region that stretched into New Jersey and Connecticut as well as up the Hudson River into New York State. However, such strategic thinking proved to be no match against the powerful political mechanisms bounding the inner city in the ubiquitous American doughnut model. The process was different in London; the famous Greater London Plan of 1944 embraced the idea of 'garden cities' and 'green belt' designated in a circle around London. Promoted as controlling urban growth, they are essentially anti-urban through their targeting the city basics of unfolding development.

Superficially, garden cities are a very attractive proposition. Originally devised by Ebenezer Howard in 1899 as an alternative to Victorian city slums, they were intended to be small urban communities containing both housing and work. Economically cohesive but small enough to also encompass a rural-like environment, this decentralizing of the city became a key component of twentieth-century planning in numerous countries. Most developed in the UK, they were integral to curtailing the growth of cities through physically bounding them by 'green belts', areas at the edge of cities wherein any urban growth was prevented. Garden cities were part of the planning beyond London's green belt. Presented as the 'lungs' of the city – open rural spaces for poor Londoners to visit but to go back home by the end of the day – green belts tended to have the same effect as American doughnut cities, with the more affluent defending the green belts that kept them separated from poorer citizens – another form of the divided city.

As a final example of misunderstanding cities the concept of the optimum size of a city mixes planning new cities with economic theory. The latter argument is that as a city grows the agglomeration externalities are initially positive but cause problems such as congestion and pollution that build up and turn the externalities negative. The point of inflexion – the point where externalities turn from positive to negative – defines the optimum size of the city. There are two obvious questions that arise from such thinking. First, the policy implications are that the state should actively engage in stopping city growth without any development replacement, which appears to be very much like the sort of anti-city policies carried out by twentieth-century communist states.

Second, such thinking represents not just another case of states undermining the autonomy of cities, but further it threatens their actual constitution as dynamic urban ecologies. There is no appreciation of the complexity of cities, no credible understanding of their nature. Ecologies in general are dynamic because they are continually solving problems of development as they arise. Thus, in the case of cities, impracticalities and problems are crucial to development; their solving is at the innovative core of what a city is and how it works. If there were no problems, that is to say a perfect urban machine, there would be no city, just a fading edifice to a misguided theory.

The destruction of cities

Being locales of wealth creation, cities have always been military targets of outside forces; they are prizes to be won, spoils of war. To defeat a city you have to cut off all its connections: siege. The offence/defence contest traditionally revolved around developing new designs of fortifications versus new siege engine technologies. There were also often rules of war whereby early surrender with payment prevented later sacking of the city. Some cities disappeared – most famously Carthage after defeat by Rome – but most cities recovered. Buildings destroyed and moveable goods plundered, yet if much of the population survived they could summon their commercial nous to regenerate agglomeration and connectivity effects: city process reactivated. But all this changed with the modern era and the development of gunpowder and cannons, changing the offence/defence contest decisively in favour of the former.

With the industrialization of war cities have had two basic roles: as the engines of war producing armaments, and as targets, latterly by bombing, to destroy production of said armaments. Thus, strategic considerations have been changed by the new circumstance that air warfare has created. The front line of the war, the specific conflict zone, is augmented by the direct bombing of cities inside the enemy's territory putting civilian citizens at risk. This is best illustrated by the contrast between the First World War, centred on trench warfare, and the Second World War with its immense bombing of cities.

The coming of aerial warfare had been ominously anticipated as a new arena for terrorizing ordinary citizens. It was expected that home morale would be undermined so the population would lose their will to continue the conflict. But this did not happen; in both Britain and Germany the bombing created a new 'war spirit' to see the conflict out to the end. That end did finally come after the dropping of atomic bombs on Hiroshima and Nagasaki in Japan but this was the result of the immense change in the level of destruction from these two single bomb drops. Thus in general aerial warfare actually adhered to traditional strategic aims of attacking cities and thereby reducing their production (agglomeration) and infrastructure (connectivity). This is not to understate the terror felt by citizens on both sides because the fundamental strategic war policies were overtly city-centric.

Cities on all sides were major victims of the Second World War; reducing cities to rubble became a commonplace outcome. In Western Europe the initial major victims were Rotterdam, London and Coventry, the last representing a relatively small city in danger of obliteration. It was followed by an extreme retribution, systematic destruction of all German cities with populations over 100,000. This devastation of city agglomerations was accompanied by systematic damage of the infrastructures connecting cities, in particular cutting off transport links between the Ruhr heavy industrial heartland from the rest of Germany.

The air offensive over Germany started in earnest in 1943 with the continuous nightly bombing of Hamburg for a week and a half. Welcome to 'area bombing', a process of systematic destruction that also went under the awful labels 'carpet bombing' and 'saturation bombing'. In Hamburg this included the creation of the first city firestorm where concentrated incendiary bombing generated its own powerful wind system to unspeakable effect. Henceforth there was a city-by-city bombing programme so that towards the end of the war there was a list of just 15 cities still left to be destroyed. One of these was Dresden. A late target, the result was a horrendous firestorm that laid low the city. All in all it left Germany with piles upon piles of rubble where there had been a system of dynamic cities, formerly some of the most productive in the world.

Beyond the goal of dislocating German war production, this area bombing encompassed a particular policy of fundamental change. This goes right back to the idea of cities as crucibles of civilization. The systematic devastation of all major German cities demolished much more than buildings, it threatened the cultural fabric in cities by obliterating all that represented German culture, vehicles for its transmission, reproduction and development – concert halls, museums, universities, schools, archives, libraries, art galleries, studios, as well as, very obviously, the architectural heritage. These are the organs that make society; they are what Jane Jacobs refers to as the 'rounding out' of the city, its human development beyond economic prowess and success. The destruction of the beautiful city of Dresden was a key marker in this process of civilization annihilation. The bombing policy – wiping German cities from the map of Europe – fitted into the infamous Morgenthau Plan that was under consideration as the war ended. This was a sort of modern Carthage solution: the idea was that the final settlement after Germany's surrender would, as well as reducing the territorial size of Germany, convert the remaining territory into a simple agricultural land; no cities meant no material basis for waging a future, third, world war.

Named after Henry Morgenthau, an influential member of the US government, for a short time his plan was well supported. However, when the reality of sustaining and rebuilding a defeated occupied country came to the fore, the plan was shelved. But it does put the area-bombing programme into a wider political context. And it does provide a very explicit recognition of cities that, although comprising only a relatively small portion of the 'national territory' of any state, are nevertheless the vital nodes in the cultural fabric of states. In fact Western Europe soon got to the far side of revenge. In 1945 the flattened German cities did contain a surviving urban population, many of whom had the knowledge and connections to create and take advantage of agglomeration and connectivity externalities and thereby turn myriad piles of rubble into dynamic cities once again. Central to the European post-war 'economic miracle', by 1960 German cities were the economic foundation of the success of the new European Common Market (later to be called the European Community).

The chapter began with the strange paradox of different treatments of cities by states in peace and war: relatively sidelined in the former, absolutely critical in the latter. But it is not quite that simple. States' security policies continue in peacetime so that a clear understanding of the importance of cities endures in this realm of state policy. For instance, in the Cold War both the USSR and the USA targeted their intercontinental ballistic weapons on each other's leading cities. This policy was called 'mutually assured destruction', 'decapitating' their rivals' societies by taking out their cities. And in the more fluid international politics of today's terrorism and counter-terrorism cities continue to be at risk from bombings by air and on land.

Concluding supplement: the right to the city

In this chapter cities have been portrayed largely as victims, once we move into the modern political realm they have very little direct power to determine their own destinies. One solution that is commonly muted is to deploy powerful mayors, thereby giving strong leadership to cities. This implies a move towards city-states as competitive political entities. But cities prosper through mutuality in networks, which does not need strong leadership but rather well-functioning city ecologies. In this argument mayors can be as superfluous as presidents and prime ministers for cultivating the latter. But their kinds of hierarchical politics are not the only way politics can be done. What is required is a very different politics, which has come to be called the right to the city.

The right to the city is a concept behind a new way of doing politics that focuses on cities. It addresses the ills of the city but without recourse to appealing to the state for solutions. It recognizes the potential of the city and thereby is able to bypass the state in its various forms – conservative, reformist or revolutionary. The key sentiment is that it is our city and we wish to reclaim it. Thus it has local, often specific, roots but is simultaneously holistic, appreciating the city in its complexity. Put simply, in an increasingly urban world there must be a better way of living in ourselves, with our family and friends, with all other people, and with the rest of nature.

Thus it is a seemingly simple claim – but with the city as its subject – being converted into a new world-changing politics. This can be summarized in five steps. (1) The language of 'rights' is usually about individuals or specific groups of individuals, but in this case the human right being claimed is collective by place: people living together in the city. (2) As such it is a claim to the urban resources around them but that are currently being appropriated by others. (3) But it goes further than this: it is a claim to make the city, to undertake its process, in a new image for a new purpose. (4) But no city exists alone, and therefore it requires making complementary links to other cities and regions. (5) Ultimately, the making of change must include ourselves, currently consuming our way to environmental oblivion; our urban way of life must be based on a new way of framing quality of life.

The right to the city has been in action as protest in twenty-first-century cities across the world as movements largely focused on the early steps listed above. The later steps are becoming increasingly urgent but, quite obviously, are very hard to implement. We will return to this need to remake our cities in the final chapter.

City Insights H: Anonymous' Berlin 1945

'Who could ever imagine such a world, hidden here, so frightened, right in the middle of a big city? This is a question asked by the anonymous author (2011) of *A Woman in Berlin*, a diary kept through the fall of Berlin to the Russian Army in 1945. She expresses this despairing thought just as Russian troops are entering her district of the city on 27 April (p. 66). This frightening new world is foreign military occupation of a large city wherein the norms and structures of the past world are banished. Everything becomes immediate in time, space, information and living: a 'timeless time ... its passing measured only by the comings and goings of men in their foreign uniforms' (p. 164), a 'horizon ... shrunk to a few hundred paces' (p. 40), all news 'from hearsay and gossip ... unclear and uncertain' (p. 135), and socially 'jumbled and scattered' people ... in patchwork households' (p. 115), a 'rampant regrouping, random alliances forming out of fear and need' (p. 120). She ends by affirming her existence: 'I only know I want to survive' (p. 308).

The diary can be divided into three phases of dreadful experiences. From 20 April to 26 April the situation is dire as the Russian troops advance. The diarist's job stops on 21 April; the German administration crumbles as water, electricity, public transport and food distribution come to an end. Citizens barricade themselves in their buildings waiting in trepidation for the enemy soldiers to appear. The first contact with soldiers comes on 27 April and a total breakdown of order ensues. Booty time – everything is plundered, both people and goods – lasts until 8 May. After 11 days of horror, military order is established involving forced labour, and the city slowly begins to come back to life as the diary ends on 22 June. The awful irony is that the German citizens are materially better off – eat better – during the booty time as the Russian troops trade plundered belongings and food supplies for alcohol and sex. Of course, it is very unequal 'trade'; let's see how it worked.

Russian troops are billeted on the street where the diarist is living with two other, older, unrelated people – a widow and a disabled man. The first few days of plunder are simply anarchic as drunken troops break into dwellings and commit mass rapes on women not able to hide. This includes the diarist and the widow. The diarist quickly devises a strategy: in the entry for 1 May she declares herself 'determined to be more than mere booty, a spoil of war' (p. 85). She speaks a little Russian, learnt from visiting

Moscow before the war, and uses this crucial skill to 'find a single wolf to keep away the pack' (p. 85). First an officer and then a major – things move quickly in this world – become her protector and 'lover' so that she can announce that, in the house where troops are always coming in and out, 'I really am taboo' (p. 103). With lots of commodities brought into the household, both liaisons appear to be a good deal, fully encouraged by the disabled man who appreciates the looted cigars. The diarist is strangely philosophical: 'should I now call myself a whore, since I am essentially living off my body, trading it for something to eat', adding only 'I'll be overjoyed to get out of this line of work' (p. 141). By the end of these brutal few days the fortitude appears commonplace.

> ... we are dealing with a collective experience ... that happened to women right and left, all somehow part of the bargain. And this mass rape is something we are overcoming collectively ... all the women help each other, by speaking about it, airing their pain and allowing others to air theirs. (p. 174)

This is reported on 8 May, a day later she announces 'there is nothing, absolutely nothing, to say about last night' (p. 182) – she had slept alone.

By 10 May a different kind of work is demanded of the women: making flags for the allies' forthcoming victory parade in the city: it seems the French flag is the easiest to make, the American the most complicated. For the first time the diarist is able to visit across town and finds that 'all around is desolation, a wasteland, not a breathe of life. This is the carcass of Berlin' (p. 191). But at the same time she notes 'One barber has reopened his shop ... The first sign of life in the city carcass' (p. 193). But the main economic activity is the deindustrialization programme, with the diarist and large numbers of other women recruited to dismantle factory machinery and load it onto trains bound for the USSR. Food rations return but uncertainty remains as she is no longer useful to the disabled man; evicted, she laments 'homeless urban nomad that I am' (p. 206).

But things are looking up. She reports: three train lines open on 14 May (p. 208); the first bus appears on 5 June (p. 284); on 8 June the S-Bahn starts running again (p. 291); as does the first tram on 13 June (p. 299). Crucially, water and electricity return. More personally, her language skills lead to her becoming an assistant in a survey of the state of banks in her part of the city: Dresdner, Commerzbank, Deutsche, the latter opened for business (p. 217). On 3 June she hooks up with an entrepreneur who is setting up a publishing company, whom we are told, 'sees skyscrapers where we see rubble' (p. 280). On 9 June she goes to the hairdressers – 'they washed about a pound of dirt out of my hair' (p. 292) – and on 13 June she sees a film at the cinema (p. 300). The publishing business is continuing to negotiate around the new bureaucracy, and beyond their venture 'other little groups of people are starting to move here and there, but in this city of islands we know nothing about each other' (p. 307). By no means back to normal, but the city is becoming a city again.

There is nothing exceptional about this episode – German troops had earlier treated Russian cities they conquered in much the same way, as all participants in this story well knew. Such has been the fate of many cities through the ages.

8. Cities globalized

Introduction: three globalizations

Cities are politically subject to the states where they are located as shown in the previous chapter. But as also noted they operate within a different social logic where complex patterns of flows – of commodities, information and people – can and do spill over state boundaries. In the last hundred years or so these myriad flows have grown immensely, resulting in our current situation being famously designated by Manuel Castells as 'network society'. Furthermore, the scale of flows has increased, with many more being global in scope, that is to say encompassing and integrating the whole world. This 'global network society' is coordinated through cities, doing work that transcends their own country's borders. For instance, today much of the financial activities that take place in London have little or nothing to do with the British economy.

What we are dealing with here are global political economies encompassing two social logics, with different functions, generating contrasting spaces. The result is a bi-layered spatial structure; a political level of coordination through states – a space of places – and an economic level operating through cities – a space of flows. This globalization of human affairs began as an expansion of European activities, both political and economic, about 500 years ago, culminating in 'global closure' – meaning all settled territories politically and economically connected – by the beginning of the twentieth century. We came across this stage in Chapter 5 as cities demanding food from a global array of supply regions. This is imperial globalization, the first of three specific globalizations distinguished by the way in which their worldwide economic connections are constituted.

This first globalization consisted of imperial politicians and traders (e.g. in Britain, France) setting the conditions for colonies (e.g. India), former colonies (e.g. in Latin America) and countries subject to unequal treaties (e.g. China) to participate in the world economy. This was coordinated through, and demanded by, three types of fast-growing cities: (a) the imperial capitals in Europe, notably London and Paris; (b) industrial cities in Europe such as Manchester; and (c) dependent cities beyond Europe dealing with the logistics such as Buenos Aires, Shanghai and Calcutta. Partially separate, a similar regional structure also developed in North America where New York functioned as the business and commercial capital complemented by industrial cities in the Midwest, notably Chicago, and with local supply cities in the west such as San Francisco, and in the south such as Atlanta. It is this specific part of imperial globalization that grew in the first half of the twentieth century to emerge as American globalization by mid-century. This second globalization heralded mass consumption to complement mass production as suburbia became the primary landscape of a new urban world, epitomized by Los Angeles. As described in Chapter 6, this new way of living diffused across the world after 1950 with the shopping mall symbolizing modern cities in the American mode worldwide.

American globalization developed into today's corporate globalization in the final quarter of the twentieth century. The third of the genre, it is the subject of this chapter. It represents a progression of Americanization but one that is increasingly shaped by other centres of economic influence, notably in Asia. The previous globalization led by US manufacturing firms with highly developed export capabilities has been superseded since the 1970s by a new suite of firms combining the computer and communication industries. Creation of near instantaneous flows of information worldwide has enabled all corporations to operate as new global strategic players. An early example was the new technology facilitating relocation of industrial production to cities in poorer countries, so as to take advantage of cheaper labour. This 'new international division of labour', as it was called, is part of the neoliberal turn in state economic policies, opening up national economies to global economic competition and thus enabling corporations to invest widely in different countries. US firms continue to be very important but are now joined by firms from many other countries, including China. It is this corporate globalization that is the highly integrated political economy that undergirds Castells' global network society wherein cities play a crucial role.

How do states and cities relate to each other within corporate globalization? In this bi-layered spatial structure the familiar international map interlocks with a world city network map. In the political layer governments control the movement of economic factors - labour and commodities – within and through their boundaries. However, these bounded spaces are not integrated functioning economies, rather they are economic jurisdictions derived from state sovereignties. It is functioning integrated economies articulated through cities that constitute the layer of flows of economic factors. Nevertheless states' economic jurisdictions are real and therefore have economic consequences. Hence lawyers in cities write trans-state contract agreements, and financial advisers provide wealth management to minimize tax liabilities. And all such firms are registered in a state and expect the protection of that state, as was noticeably evident in bailing out banks during the 2007–08 financial crisis. These are all examples of the intersection of cities and states in a bi-layered global political economy, the spaces of corporate globalization.

One final point of introduction: it is important to recognize that these three globalizations are not simply economic stages of development, rather they represent a sequence of overlapping processes with the earlier phases not disappearing but fazing into their successors, so that all are present in contemporary corporate globalization. This will become apparent in the two main sections of this chapter that deal first with the contemporary world city network wherein cities are the key nodes, and then with investigating the local forms these nodes take as multi-nodal city-regions

World city network

Corporate globalization is much more integrated than the two prior globalizations. Advances in communication technology have opened up many new practices that were simply impossible before, specifically new levels and scopes of control and organization. For instance, American companies that expanded their production into other countries in the mid-twentieth century had to operate in a decentralized manner, largely country by country: Ford executives in Detroit could not directly manage their car production in European countries. All this has changed in the network society and companies like Ford can operate genuine global strategies, with their manufacturing plants integrated worldwide as a single coherent organization.

This global shift in operational scale is not without its challenges because new practices inevitably generate new problems and thus a demand for fresh solutions. An obvious problem is that different countries have different legal regimes so that firms operating in several states require expertise on negotiating their business in multiple jurisdictions. Thus have 'global' law firms evolved to satisfy this particular new demand. If a Sydney logistics company in cooperation with a Rotterdam transport company was developing new port facilities in Cape Town financed through a bank in Frankfurt the contracts would have to be valid in Australian, Dutch, South African and German law. There are law firms, very often headquartered in London or New York, that can carry out this sort of work by having offices (and therefore 'local' expertise) in each of these cities: they will be able to bring all parts of the deal together in a single legal framing, either in English common law or New York state law. Such business services are called 'advanced producer services', firms that satisfy a range of financial, professional and creative corporate demands worldwide. As well as law firms these include advertising agencies, accountancy firms, management consultancies and numerous banking and insurance services.

These important business services existed before corporate globalization, originally plying their trade in individual cities and identified as such, for instance a New York advertising agency or an Amsterdam bank. However, as previously noted in Chapter 4, when their corporate clients expanded operations globally, the service firms had to follow or lose the business. This meant setting up new offices beyond the home city. But once in a new city there is an obvious attraction to operate in this new market rather than just work for the company they have followed. In this way corporate globalization has been responsible for many advanced producer service firms developing large office networks encompassing multiple cities. Although this globalizing process began in a haphazard way, the overall distribution of offices has been neither random nor uniform; rather, there are varying degrees of service concentrations. The results are both new economic clusters within cities and economic networks between cities.

The offices of these business service firms that form important economic clusters are often housed in the new skyscrapers that are now commonplace in cities across the world. Focusing on New York, London and Tokyo, Saskia Sassen (2001) identified a close demand–supply relationship in each of these cities: corporate headquarters provide a demand for advanced producer services and service firms are on hand to satisfy this demand. In this immensely influential work these three cities were termed 'global cities'. She envisaged only a select few of such cities; Chicago, Los Angeles, Paris, Frankfurt, Beijing, Shanghai, Hong Kong and Singapore are the other cities most commonly identified as global cities. However, there is a problem with this concept in that it implies other cities are not global in the scope of their servicing. But the network society does not work in this restrictive manner; there is no such place as a 'non-global city'. All cities are globalizing, albeit in different ways and to different degrees. Thus Castells, while embracing Sassen's global cities as a key component of his network society, extends the process to other cities, creating a much larger urban mesh penetrating economic processes through the whole of global network society. This is the world city network wherein Sassen's global cities are viewed analogously as 'tips of the iceberg' in a broader network process.

The world city network is conceived as an amalgam of leading advanced producer service firms' office networks (Taylor 2001). Every firm's office network is different, depending on their home city and the specific way they have globalized, but it will cover multiple cities overlapping greatly with the office networks of other firms. In every office the everyday work will include engagement with other offices of the firm. This is how the firm generates flows of information and knowledge between cities involving multiple tasks: instruction, assessment, strategy, advice, coordination, specialist professional inputs, tacit local nous, data analysis, planning, etc. In this way, it is firms that generate the work-flows but it is the cities that are the nodes of the world city network. Gross work-flows between cities can be estimated using data on the office networks of leading advanced producer services. Results for different years always confirm the London–New York link as the biggest link in the world city network; in fact it is the only inter-city link that actually has a name – NYLON. This reflects much more than the individual importance of each city, it implies mutuality in their relations. There is a complementarity between their respective service clusters: for instance, in financial transactions New York is the creative innovation centre and London operates as a global platform through which many of the innovations are applied and diffused. In terms of their respective commercial returns, their financial clusters generate different balances of externalities: New York provides more agglomeration advantages and London more connectivity advantages.

Some results showing the largest world city network links for 2018 based on the office networks of 175 leading advanced producer service firms are shown in Table 8.1. For ease of interpretation all links are presented as percentages of the biggest link, NYLON. The top ten links with London and New York are shown – note that for both of these two cities the largest link is with the other (i.e. constituting NYLON). The second largest link is between London and Hong Kong, just 72% the size of NYLON, further emphasizing the overall importance of NYLON in the world city network.

The first thing to note about this table is the similarity between the two lists of cities, eight cities are the same, and in the same rank order, but with the London link's percentages just slightly higher. This reflects London's function as a global platform for many firms. The two differences involve inclusion of nearby cities, Frankfurt in London's orbit and Chicago in New York's. These results indicate that at its upper echelons the world city network is highly structured, meaning that the vast majority of firms have to be in both London and New York to operate globally and very many also choose to be in the other ten cities listed.

Table 8.1 The ten largest links to London and to New York

London's main links		New York's main links	
New York	100	London	100
Hong Kong	72	Hong Kong	71
Singapore	71	Singapore	68
Paris	63	Paris	62
Shanghai	62	Shanghai	61
Beijing	61	Beijing	58
Dubai	56	Chicago	55
Tokyo	56	Tokyo	55
Sydney	56	Sydney	54
Frankfurt	52	Dubai	53

Beyond these major links, the world city network consists of myriad additional inter-city links all contributing to a continuing development of corporate globalization. The analysis from which Table 8.1 is derived includes a total of 707 cities. In studying the intricacies of this network, we find regional patterns, what we call subnets, created by firms from different parts of the world. For instance, important subnets are found in Western Europe, Latin America, South Asia and East Asia within the overall network. These are in no sense autonomous, they all relate to NYLON, but the subnets do show that the world city network is not a simple hierarchy; it features a complex geography. However, there are three large subnets that have global ranges, two existing and one emerging, which dominate the world city network.

These three subnets are termed 'extensive globalization', 'intensive globalization' and 'China globalization'. Each of them is constituted by myriad links between many cities. To make their description manageable, for each subnet I present only two sets of links focusing on just the eight leading cities for each set. The featured connections are, first, estimates of the overall links a city has within a subnet indicating their importance for the work done in that subnet, and, second, command links indicating headquarter functions, where directing the work in the subnet is located. As will be seen, the cities with most overall links in a subnet are generally not the same as the cities with the command links, but with some very notable exceptions.

Table 8.2 shows the two sets of links for the extensive globalization subnet. This is heir to imperial and early American globalization and features overall links to cities in the Global South (supply regions, now including manufacturing supply regions) serviced largely through New York. Initial US commercial expansion was into Latin America and this is clearly reflected in the overall links. Command links are very different with London joining New York at the top. The latter pairing was a key feature of American globalization with London in the role of offshore financial centre to the US economy in the immediate post-Second World War period. But in today's subnet the pattern is clearly more complex, with command links to two other US cities, three other European cities and Tokyo. The subnet is named 'extensive' for its wide range of both overall and command links.

Table 8.2   Extensive globalization subnet

Overall links	Command links
New York	New York
Bogota	London
Mexico City	Chicago
Lima	Paris
Buenos Aires	Boston
Budapest	Tokyo
Barcelona	Brussels
Prague	Dublin

Table 8.3 shows the two sets of links for the intensive globalization subnet. London now features alongside New York at the top of both lists but otherwise US cities dominate. In terms of overall links, Hong Kong and Paris are listed, the latter the only other European city to join with London. The command links list is similar except that the two cities outside the USA apart from London are Tokyo and Beijing. The way that Hong Kong and Beijing feature directly expresses their respective strategic positions: Hong Kong as long-term gateway city to China, Beijing as command centre with new global reach. This subnet echoes Sassen's global city argument; it is named intensive for its concentration of links in the major cities across the world.

Table 8.4 shows the two sets of links for the Chinese globalization subnet that has been emerging in the last decade. The list of overall links is a roll call of China's leading cities: the work done in this subnet is concentration in the one country and therefore suggests a regional-scale subnet. But the command links suggest a very different story. To be sure, China's two leading cities, Beijing and Shanghai, top the list but half of the cities featured are not Chinese. There is control of flows from New York and London once again, plus two Asian neighbours, Singapore and Tokyo. It is on the basis of the latter command links that this subnet is named a globalization subnet, albeit still emerging.

Table 8.3 Intensive globalization subnet

Overall links	Command links
New York	New York
London	London
Washington DC	Chicago
Chicago	Boston
Palo Alto	Washington DC
Los Angeles	Los Angeles
Hong Kong	Tokyo
Paris	Beijing

Table 8.4 Chinese globalization subnet

Overall links	Command links
Beijing	Beijing
Shenzhen	Shanghai
Hangzhou	Shenzhen
Shanghai	New York
Jinan	Singapore
Chengdu	Fuzhou
Kunming	Tokyo
Wuhan	London

One final point needs to be made: subnets are not like world regions with their neat, bounded patches easily depicted on a map. Rather, they are intricate patterns of links within a large whole; the intricacy is in their overlapping memberships. New York and London are the most overt example, present in all three subnets above, but such overlaps are commonplace in the very complex configuration that is the contemporary world city network.

Multi-nodal city-regions

The traditional imagery of the external relations of urban settlements has the city as a central place surrounded by its rural hinterland. Cities vary by size in direct relation to their hinterland; viewed as a hierarchy, cities were deemed to form urban systems. Before the advent of corporate globalization there was a vibrant research school studying 'national urban systems'. But this bounding of inter-city relations, not recognizing links across international boundaries such as New York–London, has been superseded by world city network analyses as described above. The focus on hinterlands, essentially local city relations, has been the remit of central place theory that emphasized hierarchical relations between urban places. Subsequently, this central place theory has been joined by central flow theory with its non-local focus as a more horizontal network structure (Taylor et al. 2010). The world city network is the contemporary expression of central flow theory. Both processes operate in all cities: 'local' competitive hierarchies within countries and 'non-local' network mutualities that transcend political boundaries. With the latter, the power of the network appears in its connectivity externalities combining with agglomeration externalities, as broadly presented by Sassen (2001) in her depiction of global city formation. Interestingly within globalizations, an important effect of this non-local process has been to produce spillover effects on nearby smaller cities.

Spillover effects generate new highly urbanized local regions that largely overshadow remaining rural parts of the hinterland. Such a process was experienced in the nineteenth century by expansion of industrial cities into industrial regions, to be labelled 'conurbations' in the early twentieth century. The spillover effects of the world city network are far greater and thereby have created much larger conglomerations of cities. These have been given various names such as 'megalopolis', 'global-city-region', 'mega-city-region' and 'polycentric metropolis'; I will call them simply multi-nodal city-regions. All the cities mentioned in the previous section as network nodes are simultaneously constituents of multi-nodal city-regions.

The study of multi-nodal city-regions has tended to be articulated through two channels: form and function. The urban form approach is associated with promotion of a new large scale of urban planning and governance. As such it is particularly concerned with maps, drawing boundaries to delimit what is to be planned. As argued in Chapter 7, such a focus on boundaries is anathema to cities as a process, turning cities into patches on maps betrays a misunderstanding of cities. In terms of function, the emphasis is on promoting cities as competitive entities, the argument is that being big is necessary to prosper in a global world. Again this is to deploy a very limited comprehension of how cities work. These criticisms are confirmed when we look at the practical proposals on offer: there is a strong emphasis on infrastructure – especially transport – with the vital importance of economic agglomeration processes linked to connectivity conspicuous by their neglect.

Time to reset, discard poorly disguised local boosterism and look at actual multi-nodal city formations that are occurring without the obstacle of a plan. The first point to make is that there is not a one-size-fits-all pattern to multi-nodal city-regions. At its simplest, these can be divided into two distinctive processes creating different types of spatial structure. The first is what comes to mind with the term spillover effect: a major city in the world city network that stimulates related work in nearby cities. The result is a primate structure, which is very familiar where one city dominates – for instance London city-region encompassing southeast England and New York city-region spreading into New Jersey and Connecticut. Certainly in these cases smaller cities benefit from being close to the two global cities. This is the general way the process called 'borrowed size' works: having 'metropolitan' products and services locally available that other cities of smaller size simply lack.

The second process of multi-nodal city-region formation is the integration of several cities of roughly the same size to form an amalgam without one dominant city. Randstad Holland is the classic case with Amsterdam, Rotterdam, the Hague and Utrecht along with many smaller cities forming a 'ring city' at the centre of the Netherlands. A key feature of such regions is that the main cities are very different, each providing diverse inputs into the regional whole. This is well illustrated in a second well-known example, the Rhine–Ruhr region in western Germany where Dortmund, Essen, Duisburg, Düsseldorf, Cologne and Bonn are the main cities. This multi-nodal city-region is a combination of long-term trading cities (Rhine) and nineteenth-century industrial cities (Ruhr). Today it is interesting for its two main cities having different connectivity profiles: Düsseldorf links more globally, Cologne more nationally. Between them they provide the region with its necessary connectivity externalities.

The four examples provided above are exemplar cases of the two processes and therefore are useful in explaining them, but they should not be read as typical of multi-nodal cities. Most multi-nodal cities have mixtures of both processes. The two multi-nodal city-regions of California illustrate this well. In Southern California the Greater Los Angeles region, including Long Beach, Anaheim, Santa Ana and Riverside, tends towards the primate structure but is no way near the primacy level of New York and London. In Northern California the San Francisco Bay Area also has a major city, a global financial centre, at its hub but here the pattern is much more diverse, with Sacramento as state political centre, Oakland/Berkeley, and San Jose in Silicon Valley where Palo Alto, the global high-tech centre, is also located. This mixture of cities creates an equally diverse cacophony of externalities, both positive and negative, where in the latter extreme housing cost differentials mark the region's material inequalities.

The final point to understand is that multi-nodal city-regions are not the static structures I have described; they are always essentially dynamic. This is best shown with a recent Chinese example, where contemporary urbanization is happening so rapidly. In South China, Guangzhou was the major beneficiary of the economic reforms after 1978. The process of primary nodal development operated with Guangzhou firmly first among a multi-nodal set of cities all manufacturing goods for export. But subsequently this hierarchical process gave way to a more horizontal development with, for instance, Shenzhen becoming a financial centre and linking with Hong Kong and Macau as a very diverse Pearl River Delta multi-nodal city-region. Like the world city network through which they operate, multi-nodal city-regions are always unfolding.

Concluding supplement: global elite enclaves

The three globalizations have produced different über-rich elites, people with a level of material wealth that divides their activities from the rest of society. In imperial globalization it was returnees from the colonies retiring to their mansions and chateaux (in late nineteenth-century USA this was the immense inequality of the so-called 'gilded age'); in American globalization it was millionaires, the 'jet set' no less; and in current corporate globalization this label has been upped to billionaires, numbered in their thousands and now being topped by the odd trillionaire. All these global elites live in the third of Castells' spaces of flows introduced in Chapter 1: enclaves of wealth where exclusive mini-networks channel the elite's material assets in a way that insulates their lives from the ordinary vicissitudes of the world economy.

Today's global elites represent the pinnacle of the immense personal inequalities enabled by corporate globalization. Their practical and cultural needs are met by an infrastructure of integrated 'islands' – some real, some gated, some priced beyond ordinary consumption – connected by private jets, yachts and blacked-out limousines. This is a very cosmopolitan elite with members from both the Global North and the Global South: entry is simply by financial worth. Its most visible institution is the annual World Economic Forum in Davos, where 'global leaders' meet in a remote skiing resort in Switzerland to put the world to rights. This has become a temporary cluster of the rich, similar to specialist annual fairs but with the specialism consisting of being very wealthy, plus invitees of the very wealthy, people who can be useful to the very wealthy. However, this large agglomeration is unusual. The enclaves of global elites are typically non-local, small agglomerations in exclusive parts of cities and fashionable leisure locales. In this situation connectivity externalities are important, both the latest electronic infrastructure and personal mini-networks, to create a sort of global-scale agglomeration of fragmented activities.

But this may be changing. The immense scale of contemporary inequalities is resulting in the development of an actual economic über-wealth agglomeration in one city, London. Rowland Atkinson (2020) identifies London as the 'alpha city', where the number of billionaires from across the world making it their 'home' (or at least one of them) has soared in recent decades. The result is that 'Wealth becomes a city industry in its own right' (p. 4). The 'industry' is the servicing of the very rich: as well as the usual suspects such as bankers (capital management), accountants (tax management) and lawyers (capital protection), there are cohorts of ordinary people brought into their economic orbit. These include people with their own power who can be useful (politicians, developers, planners) but also segments of wider services (builders, top professionals such as doctors and architects, real estate, luxury shopping, leisure pursuits, specialist art dealers and auctioneers, tailors, car dealers and many others making a living off the activities of the elite). The problem is that this is essentially not a local agglomeration: it is in London but not of London.

For instance, much of the exclusive real estate is bought but not lived in; it is merely investment, money temporarily parked until a better deal comes around. With a resultant sidelining of traditional stewardship of city and citizens, Atkinson argues that London has entered a Faustian pact with the global elite. Thus has globalization been bad for London, and increasingly dysfunctional for Londoners, despite the city's undoubted prowess in the world city network.

City Insights I: Ben Rawlence's Dadaab

Let's start with Ben Rawlence's (2016) conclusion to his book *City of Thorns* about Dadaab, a refugee camp established in 1992 in northeast Kenya to house those fleeing from the Sudanese Civil War: 'In the imagination of Somalis, even if not on the official cartography, Dadaab was now on the map' (p. 346). This is despite being located in a hostile desert, and with later opposition from the Kenyan government. Planned for less than 100,000 people, it came to house about half a million and become the largest city for 500 miles around. To understand how this happened Rawlence made numerous visits to Dadaab between 2010 and 2014 and interviewed hundreds of residents. Condensing these with specific observations of events, he provides a picture of creating livelihoods and building lives in a supposed transient settlement.

The formal situation is that the international community has responded to the plight of people fleeing Somalia by setting up a camp for refugees on land provided by the Kenyan government. A well-planned physical planning exercise provides housing and basic services in health and education alongside food aid in the form of coupons provided to registered inhabitants. However, no jobs for refugees are provided; they can volunteer to do necessary work for which they get privileges but not regular pay. All residents are expected to move on, not into Kenya, but as immigrants to Western host countries or back home to Somalia.

But reality is nothing at all like this simple picture. Dadaab has become the point of intersection for two contrasting corrupt economies. A humanitarian economy with international inputs of food sold on cheaply by refugees thereby undermining Kenyan farm production by destroying pre-existing supply and demand for food in the host country. The national winner is the southern Kenyan port of Mombasa, which handles 80% of the new food supply. At the same time there is a buoyant smuggling economy, based on sugar, coming in from the north at the Somalian port of Kismayo, feeding off tariffs, bribes and extortion, and generating huge profits for the few and multiple jobs for many more. Again Kenyan production, this time of sugar, is undercut and in decline. Thus it appears that 'those who control things that really matter in Dadaab do not live there' (p. 275). This is an elite group comprising Kenyan politicians, Somali businessmen (latterly the fundamentalist al-Shabaab) and personnel of UN organizations. Of these it appears to be the Kenyan government who is the central player. The problem is that this is 'less a state', more like 'a corrupt collection of rival cartels' (p. 135). As such, by 2014 the Kenyan government wanted to 'see Dadaab levelled' (p. 346), those now in charge viewing the city to be more trouble than it was worth, to them.

But even this more realistic interpretation is a very incomplete view: Dadaab is cursed by one-dimensional thinking. In addition to the formal and the corrupt pictures of Dadaab, there is the US State Department fitting the camp into its anti-terrorism agenda: Dadaab is seen as a breeding ground for Islamic terrorists (p. 3). This is the obverse of the al-Shabaab view of Dadaab as an opportunity for recruitment: they bring their Somalian war to Dadaab. And then there is the global media viewing Dadaab as an opportunity for a story: reporting on the latest famine and seeing yet another disaster, they refer to Dadaab as 'Haiti part two' (p. 104). All these views are important in influencing events in Dadaab but that does not mean that the people living there are helpless in the wake of all these outside influences. What Rawlence offers is a catalogue of myriad responses to changing conditions, typically generated from the outside but which are navigated by Dadaab's residents with varying degrees of success. Thus in this view, one from below, Dadaab is more than a commercial intersection, etc.; it is a complex ecology, a city process where people are actually making livelihoods and living their lives.

Dadaab is big; it is a large agglomeration and as such it provides multiple opportunities for work. There is a huge market called 'Bosnia' (named after a place that dominated the news when it was first set up) where 'you could buy everything: food, clothes, radios, generators, mobile phones, long blocks of ice' (p. 37). Bosnia is 'a cauldron of competition, struggle and uncertainty' (p. 46), a place of 'manic commerce' (p. 50). Beyond the 'bright lights' of the market there are corner shops on every block 'constructed by enterprising refugees with capital' (p. 39). There are even instances of import replacement and shifting: both pasta and ice made locally to compete with truckloads from Somalia (p. 37). Overall, the markets of Dadaab sell everything from tomatoes to trucks, with a turnover estimated as 'at least $25 million per annum' (p. 46). This is all premised on the humanitarian economy (p. 47). Thus 'emergencies' are good for the economy, a new influx of people registered with their vouchers is an opportunity (p. 90), a 'business opportunity' (p. 106). Thus there is talk of 'having a good famine' based upon 'a stampede of agencies giving out things ... liquidated for cash' (p. 163). But on the negative side, 'peace afforded only a generalised prosperity for all. The real money lay in disorder' (p. 216). For instance, a police crackdown provided 'just another licensed opportunity for extortion', protection enterprise.

But Dadaab is shown to be much more than an economic 'wild west'. There are clear signs of a rounding out of city beyond its material needs. This is largely the work of 'the '92 group', those who came as children when the camp was first opened and have therefore spent their whole lives in Dadaab (p. 149). These are the volunteers who keep the services going during emergencies when paid staff depart for their own safety. One leading resident describes his work as 'like running a medium-sized city without a budget' (p. 150). These are the people who have created a whole civil society including multiple sports teams with a particular focus on football where a complete league structure exists. Contra Kenyan government demolition policy, the fact is that 'Dadaab worked': 'through the accumulated energy of the generations that had lived there it had acquired the weight and drama of place' (p. 346). Although being 96% Somali, it has begun to become a 'cosmopolitan city' including refugees from other emergencies – Ugandans, Congolese, Burundians, Rwandans, Ethiopians – plus Kenyans attracted by the free food and services (p. 91). This is Dadaab 'on the map' (p. 346).

9. Cities in Nature

Introduction: unbalanced Natures

Cities are traditionally viewed as 'unnatural', distinct and separate from rural green landscapes where Nature flourishes for all to see and enjoy. This romantic notion is the very opposite of the treatment of cities presented in this book whereby the rural and urban constitute a single ecosystem. Cities, like everything else on Earth, are in Nature, operating as part of a global ecosystem. But cities do look somewhat different from the rest of Nature.

One fruitful way to start thinking about how cities fit into Nature is to postulate two Natures. First Nature is the Nature as we usually conceive it, the living world that exists and the environment that sustains it. Second Nature is the human uses of this world, the uses we make of First Nature to sustain and grow humanity. Cities can be seen as the pinnacle of Second Nature, the building of a complex ecology every bit as complex as First Nature. Thus it is not a coincidence that the earliest great civilizations are founded on cities exploiting the advantages of location on great rivers, First Nature constantly replenishing and enabling an 'urban revolution'. In this argument the process of the city, notably its agglomeration and connectivity advantages, is the key mechanism for creating a powerful Second Nature.

There is an obvious objection to this train of thought. All species use the environment to sustain and reproduce their kind. Do not these each constitute other species-specific Second Natures? If the global ecosystem consists of a huge functioning amalgam of 'Second Natures' what is so special about human's Second Nature? The answer is the scale at which human Second Nature operates. All other species are sustained and reproduce through their local environments, resources obtained within territories and along paths. Humans, by their trading practices, reproduce themselves through harnessing non-local resources. Developed through cities, this process has increased in size and intensity to become global. That is to say our Second Nature operates at the same scale as First Nature. It is now a danger to First Nature, most conspicuously through anthropogenic climate change and its implications for all life on Earth.

The global ecosystem has become increasingly unbalanced as our Second Nature continues to expand through its cities. This is an indictment of cities; in a very real sense, they are victims of their own success as 'growth machines'. But cities are nothing if not innovative and resourceful. The capacities of cities that created this situation are available to overcome the Natures imbalance, and these are currently being deployed to develop initiatives under the umbrella term 'green cities'. These ideas are the subject of the first section below. However, there are doubts as to the efficacy of these city-scale efforts in the face of a global-scale emergency. The second section explores the wider notion of green networks of cities as a way of scaling up to the necessary level of action. And in the final Concluding Supplement the argument of the book reaches full circle: a reckoning – cities created civilization, are they destined also to become its destroyer? Let's hope not.

Green cities

In a mischievous gibe at traditional views of cities as unnatural, David Owen (2009) has claimed New York to be the greenest community in the USA. He calls it a 'green metropolis' because its residents use less fossil fuels than anywhere else in America, including rural areas. This is the result of low car-ownership and people moving largely by public transit, made possible by the city's high residential density. I begin with this example to emphasize the point that a city being designated 'green' means contributing to human Second Nature accommodating to First Nature. Thus the Los Angeles-led horizontal city, discussed in Chapter 6 as suburbia with high consumption, will likely be greener in the literal sense – all those lovely lawns – but is the strict opposite of being a green city.

One of the key outcomes of the 1992 Earth Summit held in Rio de Janeiro was the Agenda 21 agreement that promoted local environment initiatives that could be independent of ongoing national negotiations. Supported by UN-Habitat's focus on cities, a new scale of climate change policymaking came to the fore. With national treaty-making in turns frustrating, disappointing and ultimately proving to be not effective enough, city policies have become increasingly important, sometimes countering national failings as with many US city initiatives even as the US government is abandoning international agreements. These environmental policies can be broadly divided into two types of proposed actions: mitigation to create resilient cities, and adaptation to create sustainable cities, which are dealt with in turn.

Mitigation is about assessing risks and working to contain or overcome those risks. In terms of city policies mitigation comes in two forms. First, there is disaster risk management of environmental threats. Most cities are located on low-lying ground, on coasts and rivers, and therefore are often at risk of flooding. Climate change is increasing this risk. Thus, flood control measures are a general feature of mitigation policies. More specifically in poorer cities informal housing, often on steep hillsides, is particularly vulnerable to additional extreme weather events precipitated by climate change. These are cities with large vulnerable populations whose risks range from death and spread of disease, food and clean water shortages, and general disruption of economic activities. In this situation mitigation is difficult because of weak local government and so it is more likely to come from informal initiatives by residents. These examples are all matters of resilience focused on individual cities, local initiatives to counter local threats.

The second form of mitigation derives from a general concern for risks emanating from climate change. Cities have revised their ways of dealing with a wide range of activities across their whole span of services: recycling in waste disposal, reducing leakage in water provision, insolation via housing regulations, energy efficiency in new building, low energy in public transport, and low-carbon land development. Again the means and methods vary immensely about what can be done across rich and poor cities. But in all cases prior needs of accommodating a growing population have been supplemented by new requirements to contribute to ensuring a future durable outcome. In both forms of mitigation, local and general, the goal is to produce resilient cities, able to deal with environmental threats both immediate and long term, to learn lessons from own experiences and those of other cities, and to maintain preparedness for environmental dangers.

With adaptation, policies are devised on the assumption that environmental change is inevitable. Thus this approach is much broader and more radical than mitigation; it attempts to create a sustainable city that can continue to prosper in new challenging environmental circumstances. This requires an all-encompassing vision of the city involving reduction in its environmental impact, thereby delivering for the needs of the present in ways that provide for a viable future. This involves reducing material imports such as food and energy through local production: urban agriculture and renewable energy can both be linked to waste management. Coupled with more collective use of resources – sharing appliances inside and outside the home – overall demand is reduced. And this is all helped by high residential density; the sustainable city is a compact city, encouraging walking and cycling and low energy consumption more generally.

There are a lot of overlaps in policies between these two types of green city, and most real cities will include mixes of the two sets of policies. They differ in intent and ambition. Thus, where we might agree that making a city resilient is sensible, creating a sustainable city is better characterized as well meaning. The latter evaluation comes from relating the sustainable city to the generic basics of cities that have underlain this book. Ideas about sustainable cities are dominated by architecture and planning within the remit of city governments. Thus, from Chapter 7 we understand the subject to be urban administrative areas rather than functioning city economies. Further, these governments are under increasing pressure to use high-tech data tools to become 'smart cities'. This enables more efficient energy use, for instance, and doing other services more competently, which is good. However, a collection of 'smart cities', each within their boundaries as part of a multi-nodal city-region, does not make functional sense. But the problem is much more than a boundary problem.

Smart cities are stuck in the present because they operate on data that describes the present. This is a problem with all 'evidence-based' urban policy. For instance, an origin and destination survey of car journeys in a city will indicate current bottlenecks but it will not identify supressed demand because the current road network has never linked two places.

Thus evidence will be produced to justify planning to alleviate the bottleneck but there will be no evidence to alter an entrenched demand void. This policy problem is important in developing green cities because smart tech is not a recipe for radical change. Hence if you think radical change is required to combat climate change this is not the way forward. Most cities are horizontal cities, the opposite of the compact city that likely will be part of any green urban policy agenda. And the current urban form is not a simple technical outcome – the invention of the internal combustion engine – it is a cultural outcome produced by the early twentieth-century alliance of the automotive and advertising industries that converted automobiles from rich people's toys and a menace on the roads to the icon of American modernity (McShane 1994). Thus was created a new way of urban living, now the bedrock of the super-consumer city – suburban multi-car households. It is to be hoped that we can be smart enough to subvert this car culture, but this will not simply involve making current cities specifically more efficient.

The key point is that cities are not governments, buildings, infrastructures, administrative areas or energy projects; they are agglomerations of real people making a living within a wider network realm of other cities. The resulting agglomeration and connectivity externalities generate economic development. It is this city process, generated by people's expectations and aspirations, that has to be understood in the context of climate change. It would seem that today's demanding cities can only be changed through fundamental cultural change, a reinvention of the city no less.

Green networks of cities

Let's begin by making some reasonable assumptions about what to expect looking to the future, towards the end of this century. First, there will be a lot of people, some ten billion, more or less. And most of these will be living in large cities. This outcome is not just a matter of increasing urbanization, it is simply impossible to house this many people scattered over the Earth in farms, villages and small towns, in 'green oases' of relatively self-sufficient communities. Such an idealistic vision must presume a huge cull of humanity down to a billion people more or less. So let's not go there. Staying with multiple large cities we can expect them to have increasingly green credentials – they will aspire to be resilient and sustainable, which probably means more dense cities including myriad megacities. And the latter will continue to be located in poorer parts of the world.

It is to this basic green scenario that we need to add an understanding of cities as consumption centres, forever growing, apparently out of control. This situation has come about through the basic mechanism of their complexity: growing new work to create economic development. Thus do urban dwellers make a living by taking advantage of agglomeration and connectivity externalities to generate myriad things, and in turn demand myriad things to make a life, hopefully to live a perceived better life. It is through such expectations and practices that an intricate city economy is made encompassing production and consumption, both involving supply regions – raw materials for the former, food for the latter. This economic development has been operating over millennia and has become increasingly rapid in recent centuries. Climate change is telling us it cannot continue. Cities have become too successful for their own good, meaning us residents producing and consuming like there is no tomorrow. Our city-centred future simply cannot be a projection of the present: the Earth is not big enough.

Let's start again: cities are where economic supply and demand plays out *in extremis*. But cities are human's greatest invention, they are where our creativity blossoms so that we have a specifically dynamic Second Nature, a vitality and dynamism we should strive, not just to maintain, but to nurture and progress, albeit in new ways. This means growing new work based upon city externalities but with the resulting dynamism creating a different economics to one that costs the environment. In this way, the current imbalance between First Nature and our Second Nature can begin to be addressed explicitly through cities; people making a living to make a life for themselves and for future generations.

What we are searching for is a conversion of the city process whereby the advantages of agglomeration and connectivity are deployed in new ways. Mitigation to build resilient cities, yes; adaptation to build sustainable cities, yes – these are both necessary, but they are also insufficient through their not incorporating the nature of cities as incessant development. The essential augmentation of the city process is stewardship to build posterity cities. This is a much more holistic vision of cities, not a solution to specific urban problems but an all-encompassing reinvention of cities. Work in cities will still grow cities but the growth has a different dynamism of individual and collective stewardship aimed at balancing First and Second Natures.

Posterity cities are super-stewardship centres. First and foremost they are cities and therefore the characteristics described in the substantive chapters of this book will have to continue. The question is how will they each contribute to a transition from heady consumption to meaningful stewardship?

- Busy cities remain hives of immense activities in general as agglomeration, and with new specialist clusters. Stewardship requires a boom in creativity. The goals of new work will have to be reconfigured and this means a new rewards mechanism that fully costs the environmental impact of resulting changes. Network society requires less physical contact, but myriad face-to-face interaction is still necessary to foster 'accidental' links across economic sectors, a crucial source of change in all cities.
- Cities will have to remain connected to prevent the ossification of isolation. Stewardship has to be cosmopolitan to foster diffusion and mixing of ideas. But there will also have to be movement of physical items, trading of necessary raw materials and food production that are unevenly spread across the Earth's surface. There will have to be a complete reappraisal of local and non-local.
- Cities continue to be demanding, but in different ways. They will require intensive efforts to overcome their problems emanating from climate change, substituting old work for new work that is stewardship. Cities are at the heart of the problems ahead and it is here that challenges are confronted and creatively turned into stewardship work, both cerebral and physical. These are new ways of making a living to make a life, the basic demand on all cities.
- Cities will remain divided because migration will continue. The form of stewardship will vary across the world – the idea that one stewardship approach fits all invites stagnation for all. But the key point here is that Nature and its stewardship includes people. The inter-city flows that today maintain material inequalities between rich cities and poor cities will be reconfigured to equalize values between cities so as to sustain and nurture human life. Difference has to be celebrated as essential to stewardship.

- The relationship between cities and states will have to be thoroughly reformed to enable posterity cities. There will continue to be a need for rules and order, but the rigid political space of places based upon absolute sovereignties is simply not fit for purpose in world stewardship. This is possibly the severest challenge to stewardship because the inside/outside boundary mechanism that states operate by, emphatically endorsed by nationalisms, is anathema to wider stewardship via the innate porosity of cities.
- Cities will continue to be globalized for trading reasons mentioned previously but also because all contributions from across the world are required to build a complex stewardship of multiple posterity cities, each with its own individuality and therefore making its special unique contribution to fostering both First and Second Natures. Differences between cities are the first requirement for dynamic innovative networks: today's world city network will have to be transformed into multiple green networks of cities.

These six glimpses of transitions are all immensely speculative. There is no static utopia to aim for. Rather there are principles deriving from the notion of stewardship that appear to be necessary for creating posterity cities. But the latter is not a goal as such, but a continuing moving set of city processes with variable outcomes over space and time. There can be no model, just multiple movements.

However, we can think about the rolling outcome of stewardships in dynamic terms as a geographical imagination. Current concern for resilient cities and sustainable cities shows limited consideration of the world outside a particular city. There is the concept of ecological footprint that can be applied to cities, indicating the relative impact of a city on the environment. New York might be cast as the greenest community in the USA by energy consumption but it is a long, long way from feeding itself! But such a footprint measure hardly constitutes a geographical imagination of a future world. All city-by-city studies imply a grossly simple geography of a landscape dotted with resilient and/or sustainable cities. There is no logic that suggests cities have been, are, or ever could be, just a scattering of settlements, green or otherwise, in a functionless economic void. In complete contrast, multiple green networks of cities straddling the Earth do give us a coherent geographical imagination that makes sense, something to think about: a possibility of posterity cities as a green network society.

Concluding supplement: cities reckoning?

Negotiations between states on combating climate change have always been difficult and have come to be widely viewed as failing. Hence the argument above that states, with their inherent competitiveness, are not fit for this purpose. It might follow therefore that cities, with their tendency to complementarity, are the more likely human vehicle to create a common front for the climate challenge. That has been the underlying assumption above in deriving the concept of green networks of cities: the social mechanism behind climate change has been economic rather than political. Ultimately, it has been ever-growing city demand – super-consumption – that has precipitated the environmental conditions expressed as climate warming. Thus, it is argued, solutions have to be sought through cities. But what if the timetable for the social changes needed, as minimally presented above, is out of sync with a more rapid environmental timeline of changes?

The answer to the question is that the story of cities will have come full circle: a causal path from the birth of civilizations to the end of civilization. If unbalanced Natures are not brought into some sort of equilibrium it is Second Nature that will be the loser. This would be a reckoning for cities, a retribution for uncontrolled selfish creativity. To survive in Nature species are required to accomplish two things: biological reproduction of individuals, and ecological reproduction of their supporting environment. Through cities we are falling short on the second reproduction.

This cities reckoning argument counters the tendency throughout this book to emphasize positivity in cities relative to the many downsides: the experiences of myriad individuals facing the huge economic leviathan, city as destroyer of hopes and lives. In riposte, all I can say is that futures decades away can never be known, in either positive or negative terms. That is why the green networks of cities were framed as a geographical imagination without any of the detail expected of model predictions. Yes, we cannot know, but we can care, and a working knowledge of cities will be indispensable for converting care into action if and when the time is right.

Bibliographic notes and references

Full references for *City Insights*

If the reader finds one or more of the *City Insights* particularly compelling, here are the full citations so that they can be followed up.

Anonymous (2011) *A Woman in Berlin: Diary 20 April 1945 to 22 June 1945*. London: Virago.

Brook, T. (2008) *Vermeer's Hat: The Seventeenth Century and the Dawn of the Global World*. New York: Bloomsbury.

Cronon, W. (1991) *Nature's Metropolis: Chicago and the Great West*. New York: Norton.

Hassett, B. (2017) *Built on Bones: 15,000 Years of Urban Life and Death*. London: Bloomsbury.

Lloyd, T. H. (1991) *England and the German Hanse 1157–1611: A Study of Their Trade and Commercial Diplomacy*. Cambridge: Cambridge University Press.

Navai, R. (2014) *City of Lies: Love, Sex, Death and the Search for Truth in Tehran*. London: Weidenfeld & Nicolson.

Pai, H.-H. (2012) *Scattered Sand: The Story of China's Rural Migrants*. London Verso.

Rawlence, B. (2016) *City of Thorns: Nine Lives in the World's Largest Refugee Camp*. London: Portobello.

Soares, L. E. (2016) *Rio de Janeiro: Extreme City*. London: Allen Lane.

References cited in the text

References in the text have been kept to a minimum in an effort to make reading easier. Thus, they have been used only when an idea or topic is drawn explicitly from an author, who should thus be cited as the source.

Algaze, G. (2005a) *The Uruk World System: The Dynamics of Expansion of Early Mesopotamian Civilization*. Chicago, IL: University of Chicago Press.
Algaze, G. (2005b) 'The Sumerian take-off', *Structure and Dynamics: eJournal of Anthropological and Related Sciences* 1 (1), Article 2.
Amin A. and N. Thrift (2017) *Seeing Like a City*. Cambridge: Polity.
Atkinson, R. (2020) *Alpha City: How London Was Captured by the Super-Rich*. London: Verso.
Batty, M. (2013) *The New Science of Cities*. Cambridge, MA: The MIT Press.
Burgess, E. W. (1925) 'The growth of the city', in R. E. Park, E. W. Burgess and R. D. McKenzie (eds) *The City*. Chicago, IL: University of Chicago Press.
Castells, M. (1996) *The Rise of Network Society*. Oxford: Blackwell.
Childe, V. G. (1950) 'The urban revolution', *Town Planning Review* 21, 3–17.
Clement, C. R., W. M. Denevan, M. J. Heckenberger, A. B. Junqueira, E. G. Neves, W. C. Teixeira and W. I. Woods (2015) 'The domestication of Amazonia before European conquest, *Proceedings B, Royal Society* (http://rspb.royalsocietypublishing.org/ (accessed 29 September 2015)).
Davies, N. and R. Moorhouse (2003) *Microcosm: Portrait of a Central European City*. London: Pimlico.
Dearlove, J. (1979) *The Reorganization of British Local Government: Old Orthodoxies and a Political Perspective*. Cambridge: Cambridge University Press.
Glaeser, E. L. (2011) *The Triumph of the City*. London: Macmillan.
Glaeser, E. L., H. D. Kalial, J. A. Scheinkman and A. Schleifer (1992) 'Growth in cities', *Journal of Political Economy* 100, 1126–52.
Gregory, D. (1982) *Regional Transformation and Industrial Revolution: A Geography of the Yorkshire Woollen Industry*. London: Macmillan.
Jacobs, J. (1970) *The Economy of Cities*. New York: Vintage.
Jacobs, J. (1984) *Cities and the Wealth of Nations*. New York: Vintage.
Jacobs, J. (2000) *The Nature of Economies*. New York: Vintage.
Mann, C. C. (2011) *1491: New Revelations of the Americas Before Columbus*. New York: Vintage.
Markusen, A. (1996) 'Sticky places in slippery space: a typology of industrial districts', *Economic Geography* 72, 293–313.
Marshall, A. (1890) *Principles of Economics*. London: Macmillan.
McShane, C. (1994) *Down the Asphalt Path: The Automobile and the American City*. New York: Columbia University Press.
Modelski, G. (2003) *World Cities –3000 to 2000*. Washington, DC: Faros 2000.
Nissen, H. J. (1988) *The Early History of the Ancient Near East, 9000–2000 BC*. Chicago, IL: University of Chicago Press.
Owen, D. (2009) *Green Metropolis*. New York: Riverhead.
Sassen, S. (2001) *The Global City: New York, London, Tokyo*. Princeton, NJ: Princeton University Press.

Scott, A. J. (2008) *Social Economy of the Metropolis: Cognitive-Cultural Capital and the Global Resurgence of Cities*. Oxford: Oxford University Press.
Scott, A. J. and M. Storper (2014) 'The nature of cities: the scope and limits of urban theory', *International Journal of Urban and Regional Research* 39, 1–15.
Smith, M. E. (2016) 'How can archaeologists identify early cities? Definitions, types and attributes', in M. Fernandez-Gotz and D. Krausse (eds) *Eurasia at the Dawn of History: Urbanization and Social Change*. Cambridge: Cambridge University Press.
Taylor, P. J. (2001) 'Specification of the world city network', *Geographical Analysis* 33, 181–93.
Taylor, P. J., M. Hoyler and R. Verbruggen (2010) 'External urban relational process: introducing central flow theory to complement central place theory', *Urban Studies* 47, 2803–18.
Weber, A. F. (1899) *The Growth of Cities in the Nineteenth Century: A Study in Statistics*. Ithaca, NY: Cornell University Press.
Wilson, G. (1971) *Gentleman Merchants: The Merchant's Community in Leeds, 1700–1830*. Manchester: Manchester University Press.
Wirth, L. (1938) 'Urbanism as a way of life', *American Journal of Sociology* 44, 3–24.

Recent books that take an overall view of cities

These are books that develop further, and in different ways, many of the arguments developed above. These include five books that appear in the previous list – Amin and Thrift (2017), Batty (2013), Glaeser (2011), Jacobs (2000) and Sassen (2001) – and the following titles.

Brenner, N. (ed.) (2014) *Implosions/Explosions: Towards a Study of Planetary Urbanization*. Berlin: Jovis.
Glaeser, E., K. Kourtit and P. Nijkamp (eds) (2020) *Urban Empires: Cities as Global Rulers in the New Urban World*. London: Routledge.
Hall, P. (1988) *Cities in Civilization*. London: Weidenfeld & Nicolson.
Perulli, P. (2017) *The Urban Contract: Community, Governance and Capitalism*. London: Routledge.
Robinson, J. (2006) *Ordinary Cities: Between Modernity and Development*. London: Routledge.
Scott, A. G. (2012) *A World in Emergence: Cities and Regions in the 21st Century*. Cheltenham, UK and Northampton, MA, USA: Edward Elgar.
Soja, E. W. (2000) *Postmetropolis: Critical Studies of Cities and Regions*. Oxford: Blackwell.
Storper, M. (2013) *Keys to the City: How Economics, Institutions, Social Interaction and Politics Shape Development*. Princeton, NJ: Princeton University Press.
Taylor, P. J. (2013) *Extraordinary Cities: Millennia of Moral Syndromes, World-Systems and City/State Relations*. Cheltenham, UK and Northampton, MA, USA: Edward Elgar.

Publications that further inform chapter subjects

These are either publications that I know well and are therefore implicit in the text of some subjects, and/or are particularly suitable to begin reading beyond my text on a subject. I have limited the list to just three references per substantive chapter.

Cities as the birth of civilizations

Smith, M. L. (ed.) (2003) *The Social Construction of Ancient Cities*. Washington, DC: Smithsonian Books.

Thomas, A. R. (2010) *The Evolution of the Ancient City*. Lanham, MD: Rowman & Littlefield.

Yoffee, N. (2005) *Myths of the Archaic State: Evolution of the Earliest Cities, States, and Civilizations*. Cambridge: Cambridge University Press.

Busy cities

Fujita, M. and J.-F. Thisse (2002) *Economics of Agglomeration: Cities, Industrial Location and Regional Growth*. Cambridge: Cambridge University Press.

Kratke, S. (2011) *The Creative Capital of Cities: Interactive Knowledge Creation and the Urbanization Economics of Innovation*. Oxford: Wiley-Blackwell.

Porter, M. E. (1998) 'Clusters and the new economics of competition', *Harvard Business Review* 73 (3) 55–71.

Cities connected

LaBianca, O. S. and S. A. Scham (eds) (2010) *Connectivity in Antiquity: Globalization as Long-term Historical Process*. London: Equinox.

Neal, Z. P. (2013) *The Connected City: How Networks Are Shaping the Modern Metropolis*. London: Routledge.

Sassen, S. (2006) *Cities in a World Economy*. Thousand Oaks, CA: Pine Forge.

Demanding cities

Arrighi, G. (2010) *The Long Twentieth Century: Money, Power and the Origins of Our Times*. London: Verso.

Bowden, C. (2011) *Murder City: Ciudad Juarez and the Global Economy's New Killing Fields*. New York: Nation Books.

Steel, C. (2008) *Hungry City: How Food Shapes Our Lives*. London: Chatto and Windus.

Divided cities

Davis, M. (2006) *Planet of Slums*. London: Verso.
Hamnett, C. (2003) *Unequal City: London in the Global Arena*. London: Routledge.
Knox, P. L. (2008) *Metroburbia, USA*. New Brunswick, NJ: Rutgers University Press.

Cities in states

Grayling, A. C. (2006) *Among the Dead Cities: Is the Targeting of Civilians in War Ever 'Justified?'*. London: Bloomsbury.
Hall, P. (1996) *Cities of Tomorrow: An Intellectual History of Urban Planning and Design in the Twentieth Century*. Oxford: Blackwell.
Harvey, D. (2012) *Rebel Cities: From the Right to the City to the Urban Revolution*. London: Verso.

Cities globalized

Harrison, J. and M. Hoyler (eds) (2015) *Megaregions: Globalization's New Urban Form?* Cheltenham, UK and Northampton, MA, USA: Edward Elgar.
Lizieri, C. (2009) *Towers of Capital: Office Markets and International Financial Services*. Oxford: Wiley-Blackwell.
Taylor, P. J. and B. Derudder (2016) *World City Network: A Global Urban Analysis*. London: Routledge.

Cities in Nature

Bulkeley, H. (2013) *Cities and Climate Change*. London: Routledge.
Girardet, H. (2008) *Cities, People, Planet*. Chichester: John Wiley.
Taylor, P. J., G. O'Brien and P. O'Keefe (2020) *Cities Demanding the Earth: A New Understanding of the Climate Emergency*. Bristol: Bristol University Press.

Publications of large collections of papers

One of the reactions to the publication boom described in the Preamble has been the production of large compilations of influential papers or commissioned chapters. The four listed below include over 250 chapters for the reader to peruse and develop knowledge beyond this book's introduction to cities.

Derudder, B., M. Hoyler, P. J. Taylor and F. Witlox (eds) (2012) *International Handbook of Globalization and World Cities*. Cheltenham, UK and Northampton, MA, USA: Edward Elgar.
LeGates, R. T. and F. Stout (eds) (2016) *The City Reader* (6th edition). London: Routledge.
Norwich, J. J. (eds) (2009) *Great Cities in History*. London: Thames & Hudson.
Ren, X. and R. Keil (eds) (2018) *The Globalizing Cities Reader* (2nd edition). London: Routledge.

Index

adaptation policies 117, 119
Addis Ababa 56
advanced producer services 100–102
agglomeration 6–8, 22, 51–2
    city networks 119
    civilizations 28–9
    connectivity and 10, 60
    dependency and 66
    eclectic effects 43–5
    economic demand 63
    elite enclaves 109
    explosive city growth 54–5
    networking 49
    power and 11
    sectorial clusters 39–43
    state relations 12, 84
agglomeration externalities 7–9, 13–15, 39–40, 56–7, 88
Akkad city 30–31
Algaze, Guillermo 28–9
Amazonian cities 27, 31–3
American civil war 10
American globalization 98, 103, 108
Amin, A. 4
archaeology 4, 26–7, 31, 33
Atkinson, Rowland 109–10

Batty, M. 4
Beijing 63, 104
Berlin occupation 94–6
bioarchaeology 35–7
bombing cities 44, 89–91
borrowed size process 42, 107
Boston, England 58–60
bounding cities 14, 18, 85, 99, 106, 121
Brazil 47–8

Britain, Second World War 89–90
Brook, Timothy, *Vermeer's Hat* 23–5
Burgess, Ernest 74
business services 52–3, 100–101

Californian city-regions 108
capital cities 12–13, 44, 78
car use survey 117–18
Castells, Manuel 8–9, 50–51, 97–8, 101, 109
cathedral cities 2
central business districts (CBDs) 74
central flow theory 106
central place theory 106
Champagne Fairs 62–3
Chicago 69–71, 76–7
Childe, Gordon 4
China
    Delft connection 23–5
    megacities 45
    multi-nodal city-regions 108
    revolutions 86
    rural migrants 80–82
China globalization 103–5
cities
    busy-ness 38–48, 120
    civilizations and 26–37, 90, 114–15, 122
    concept of 17–18
    criteria 4
    debasing 85–9
    definitions 1, 36
    diversity 43
    divisions within 72–82, 120
    domination 13–14
    growth 18, 54–6, 88

language differences 2–3
modern paradox of 83–4
specifics 15–17
in states 30–31, 83–96
city agglomeration *see* agglomeration...
'city boundaries' 85
city demand 61–72, 101, 120
city generics 2–17, 19, 55
city networks 58–60, 118–22
   see also world city network
city-regions 41–2, 87, 106–8
city-states 29–31, 92
civilizations 26–37, 90, 114–15, 122
Clement, C. R. 31–2
climate change 34, 115–16, 118–20
cluster arrangement types 41
command links 103–4
command power 10–14
commercial cities 29–30
commodity flows 69
concentric model, city structure 74
connectivity 8–10, 49–60, 63, 84, 119
connectivity externalities 56–7
consumption 39, 61–4, 66, 74–6, 98, 119
'conurbations' 87, 106
core–periphery approach 65
corporate globalization 51–2, 75, 98–100, 103, 108–9
Covid-19 pandemic 37
criminality 44
Cronon, William, *Nature's Metropolis* 69–71
cultural assimilation 77
cultural cosmopolitanism 76–8
cultural diaspora 77–8
cultural segregation 76–7

Dadaab city 111–13
Davies, N. 14
Delft 23–5
demand from cities 61–71, 72, 101, 120
demography, limitations 1
dependency 11, 64–7, 98
destruction effects 89–91
Detroit 40
diseases 35–7

Dresden, bombing of 90–91
dual economic development 66
dynamism of cities 1, 15, 88, 119–20

ecological dynamism 88
ecological footprint concept 121
ecological power 11–12
economic clusters 40
economic demand 62–3, 64, 67
economic development 56, 83, 119
economic externalities 41–3
economic inequalities 61, 72–3, 75–6
economic jurisdictions 99
economic specialization 40
economic spurts 54–5, 62
economies, cities as 4–5, 38–9
elite power 47
elite separation 9, 108–12
employment growth 40–41
England
   German Hanse in 58–60
   Industrial Revolution 56–7
   local government reform 86–7
entertainment industry 43
epidemics 35
explosive city growth 54–6
extensive globalization 103–4
externalities, definition 6

fairs 62–4
favelas 47–8
financial centre clusters 11
flood control 116
Florence city 83–4, 85
food demand 63–4
foreign occupations 44, 94–6
French Revolution 85
functionalism 86–7

garden cities 88
gateway cities 70
Geddes, Patrick 87
generic thinking 2–17, 19, 55
gentrification 75–6
German Hanse 58–60
Germany
   economic spurts 54

occupation of Berlin 94–6
   Second World War 89–91
ghettoes 77
Glaeser, Edward 6, 40–41
global cities 26, 51–3, 68, 100–101
global ecosystem 114–15
global elite enclaves 108–10
global network society 9, 98
global networking, examples 53
globalization 97–113
   corporate organization 75
   disease and 37
   dual economic development 66
   gentrification 76
   posterity cities 121
green belts 88
green cities 115–18, 121
green networks 118–21, 122

Hanse network 58–60
Hassett, Brenna, *Built on Bones* 35–7
high-tech clusters 42
hinterland–city relation 11, 70–71, 106
Hollywood 7–8, 40
Hong Kong 12, 104
Howard, Ebenezer 88

imperial globalization 97–8, 103, 108
import replacement 55–6, 60, 66
import shifting 55, 60, 66
industrial agglomerations 57
industrial cities 51, 66, 73, 79
industrial clusters 7, 40, 41
industrial districts 4, 39, 41–2
'industrial farming' 66
Industrial Revolution 56–7, 64
inequalities
   corporate economy 75
   creating 61
   cultural segregation 76
   inter-city flows 120
   migration and 72–3
infrastructures 8–9
institutions 2–3
instrumental functionalism 86–7
intensive globalization 103–5
inter-city flows 120
inter-city mutuality 65

Iranian cities 21–3
Islamic cities 21–3

Jacobs externalities 40, 43, 54
Jacobs, Jane 5, 40, 41, 54–6, 90
Jingdezhen 24–5

Kampuchea revolution 86
Kenyan refugees 111–13
kingship 29
kontors 58–9

labour costs, relocations 98
labour demand 72
labour demonstrations 78–9
labour inequalities 73
land control 14
land speculators 69
landscapes 31–2, 61–2, 67
law firms 100
Lloyd, T. H., *England and the German Hanse 1157–1611* 58–60
local government reforms 86–7
logistics hubs 65
London
   city-regions 107
   global elite enclaves 109–10
   network externality 8, 9
   population size 64
   regional designation 87–8
   Steelyard kontor 58–60
   underground system 6
   world city network 101–5
Los Angeles 40, 54, 67, 115
   see also Hollywood
Lula da Silva, Luiz Inácio 47–8

Manchester 67, 78–9
Mann, C. C. 32
markets
   consumption 62–4
   Dadaab market 112
   demand from cities 61
   monoculture 66
Marshall, Alfred 4–5, 39, 41–2
mass consumption 98
material inequalities 76, 120

meat industry 70
megacities 45–6, 119
Mesopotamian cities 2, 26–31, 61
metro systems 6
Michelangelo's *David* 83–4
migration 44, 72–3, 76–7, 80–82, 120
Milan 10
mitigation policies 116–17, 119
Modelski, George 27, 30
'modern cities' 66
monoculture 66
Moorhouse, R. 14
Morgenthau, Henry 91
Morgenthau Plan 91
multi-nodal city-regions 42, 106–8

naming rights to cities 13, 14
nature 114–22
Navai, Ramita, *City of Lies* 21–3
network of cities
    creating 49–53
    German Hanse network 58–60
    nature and 118–21, 122
network externalities 8–9, 13–14, 15, 28
network society 51, 97–105, 120
networked power 11–12
new work 38–9, 72, 74–5, 119–20
New York
    city-regions 107
    as green city 115, 121
    network externality 8
    regional designation 87
    transport systems 6
    world city network 101–5
Nissen, H. J. 30

old work 38–9, 66, 120
optimum size concept 88
organized complexity 5–6, 11–12
Owen, David 115

Pai, Hsiao-Hung, *Scattered Sand* 80–82
pandemics 36–7
Paris, France 10, 78, 85
plant domestication 32

political economy 97–9
political potential, city-states 92
political power 83–4
political protest 44
political revolutions 78–9, 86
pollution 68
population size
    city demand 64
    dynamic cities 1
    megacities 45–6
    urbanism 4
positionality concerns 19–20
posterity cities 119–21
power
    of economic demand 67
    of the elite 47
    political 83–4
    projection of 10–12
    see also command power
pre-Columbian Amazonian civilizations 27, 31–3
process, cities as 5–6, 36
production clusters 39–40
provinces 30
public sector 44

Randstad cities 42, 107
Rawlence, Ben, *City of Thorns* 111–13
refugee camps 111–13
regional designation of cities 87–8
regional economies 64
reification 19
residential segregation 77
resilient cities 116–17, 118–19, 121
retail sector 43
revolutions 78–9, 85–6
right to the city 92–3
Rio de Janeiro 47–8
risk assessments 116
rivers, city emergence link 27, 33, 50, 114
roadways 50
Roman cities 35–6
Rome, landscape making 63
rural, use of term 67–8
rural–city relation 71
rural hinterlands 106
rural landscapes 67, 114

rural–urban migration 73, 80–82
rush hours 38
Russian occupation, Berlin 94–6

San Francisco 42
Sargon the Great 30
Sassen, Saskia 52, 101, 106
Scott, Allen J. 4, 7
Second World War 89–91
sectorial clusters 39–43
security 85–6
shopping centres 39, 98
skyscrapers 75, 100–101
slums 73, 75
smart cities 117–18
Smith, M. E. 4
Soares, Luiz Eduardo, *Rio de Janeiro* 47–8
social change 10–12
social interactions 9
social space 8
Somalian refugees 111–13
South African segregation 77
spaces of flows 8–9, 17–18, 50, 97, 109
spaces of places 8, 97
spatial structures 73–6, 97
specialist clusters 40–41, 65
specialization dependency 64–7
spillover effects 106–7
sports industry 43
state–city relations 12–15, 25, 36, 79, 83–96, 121
states
    civilization and 27–31
    climate change and 122
    domination 13
    economic jurisdictions 99

stewardship 110, 119–21
Storper, M. 4
structural power 11
suburbanization 74–5, 87, 98
Sumerian cities 27–31
supply–demand processes 61, 101
supply regions 64–7, 119
sustainable cities 116, 117, 119, 121

Tehran 21–3
Thrift, N. 4
Tokyo 55
tourism 48, 81–2
trading mechanisms 62–3
transport systems 6, 50
Turner, Frederick Jackson 69

underground system, London 6
Ur city 27, 30
urban, concept of 17–18
urban form approach 106–7
urban function approach 107
urban societies' rural land 67–8
urbanism, definition 4
urbanization 31–3, 35
Uruk city 27, 29–30

Vermeer, Johannes 23–5

war 44, 84, 89–91
water shortages 67
Wirth, Louis 4
A Woman in Berlin (anon.) 94–6
work-flows 101
world city network 51–3, 99–106
writing, invention of 28–9
Wroclaw example 14

Titles in the **Elgar Advanced Introductions** series include:

International Political Economy
Benjamin J. Cohen

The Austrian School of Economics
Randall G. Holcombe

Cultural Economics
Ruth Towse

Law and Development
Michael J. Trebilcock and Mariana Mota Prado

International Humanitarian Law
Robert Kolb

International Trade Law
Michael J. Trebilcock

Post Keynesian Economics
J.E. King

International Intellectual Property
Susy Frankel and Daniel J. Gervais

Public Management and Administration
Christopher Pollitt

Organised Crime
Leslie Holmes

Nationalism
Liah Greenfeld

Social Policy
Daniel Béland and Rianne Mahon

Globalisation
Jonathan Michie

Entrepreneurial Finance
Hans Landström

International Conflict and Security Law
Nigel D. White

Comparative Constitutional Law
Mark Tushnet

International Human Rights Law
Dinah L. Shelton

Entrepreneurship
Robert D. Hisrich

International Tax Law
Reuven S. Avi-Yonah

Public Policy
B. Guy Peters

The Law of International Organizations
Jan Klabbers

International Environmental Law
Ellen Hey

International Sales Law
Clayton P. Gillette

Corporate Venturing
Robert D. Hisrich

Public Choice
Randall G. Holcombe

Private Law
Jan M. Smits

Consumer Behavior Analysis
Gordon Foxall

Behavioral Economics
John F. Tomer

Cost-Benefit Analysis
Robert J. Brent

Environmental Impact Assessment
Angus Morrison Saunders

Comparative Constitutional Law
Second Edition
Mark Tushnet

National Innovation Systems
Cristina Chaminade, Bengt-Åke Lundvall and Shagufta Haneef

Ecological Economics
Matthias Ruth

Private International Law and Procedure
Peter Hay

Freedom of Expression
Mark Tushnet

Law and Globalisation
Jaakko Husa

Regional Innovation Systems
Bjørn T. Asheim, Arne Isaksen and Michaela Trippl

International Political Economy
Second Edition
Benjamin J. Cohen

International Tax Law
Second Edition
Reuven S. Avi-Yonah

Social Innovation
Frank Moulaert and Diana MacCallum

The Creative City
Charles Landry

European Union Law
Jacques Ziller

Planning Theory
Robert A. Beauregard

Tourism Destination Management
Chris Ryan

International Investment Law
August Reinisch

Sustainable Tourism
David Weaver

Austrian School of Economics
Second Edition
Randall G. Holcombe

U.S. Criminal Procedure
Christopher Slobogin

Platform Economics
Robin Mansell and W. Edward Steinmueller

Public Finance
Vito Tanzi

Feminist Economics
Joyce P. Jacobsen

Human Dignity and Law
James R. May and Erin Daly

Space Law
Frans G. von der Dunk

Legal Research Methods
Ernst Hirsch Ballin

National Accounting
John M. Hartwick

International Human Rights Law
Second Edition
Dinah L. Shelton

Privacy Law
Megan Richardson

Law and Artificial Intelligence
*Woodrow Barfield and
Ugo Pagello*

Politics of International
Human Rights
David P. Forsythe

Community-based Conservation
Fikret Berkes

Global Production Networks
Neil M. Coe

Mental Health Law
Michael L. Perlin

Law and Literature
Peter Goodrich

Creative Industries
John Hartley

Global Administration Law
Sabino Cassese

Housing Studies
William A.V. Clark

Public Policy
B. Guy Peters

Global Sports Law
Stephen F. Ross

Empirical Legal Research
Herbert M. Kritzer

Cities
Peter J. Taylor

Law and Entrepreneurship
Shubha Ghosh

Mobilities
Mimi Sheller

Technology Policy
*Albert N. Link and James
Cunningham*

Urban Transport Planning
Kevin J. Krizek and David A. King